Biodiversity and Conservation in Forests

Biodiversity and Conservation in Forests

Special Issue Editor

Diana F. Tomback

MDPI • Basel • Beijing • Wuhan • Barcelona • Belgrade

MDPI

Special Issue Editor
Diana F. Tomback
University of Colorado Denver
USA

Editorial Office
MDPI
St. Alban-Anlage 66
Basel, Switzerland

This is a reprint of articles from the Special Issue published online in the open access journal *Forests* (ISSN 1999-4907) from 2017 to 2018 (available at: http://www.mdpi.com/journal/forests/special_ issues/biodiversity_conservation)

For citation purposes, cite each article independently as indicated on the article page online and as indicated below:

LastName, A.A.; LastName, B.B.; LastName, C.C. Article Title. *Journal Name* **Year**, *Article Number, Page Range.*

ISBN 978-3-03897-574-8 (Pbk)
ISBN 978-3-03897-575-5 (PDF)

Cover image courtesy of Diana F. Tomback.

Contents

About the Special Issue Editor

Diana F. Tomback is Professor in the Department of Integrative Biology at the University of Colorado Denver. Her expertise includes evolutionary ecology, with application to forest ecology and conservation biology. She is known for her studies of Clark's nutcracker, a bird of high mountain forests, and its interaction with several white pine species, especially whitebark pine (*Pinus albicaulis*), leading to her election in 1994 as Fellow of the American Ornithologists' Union. Her research over time has revealed major ecological and evolutionary consequences to pines from avian seed dispersal, including growth form, population structure, regeneration biology, and the effects of exotic disease and mountain pine beetles on the bird-pine mutualism. Tomback was lead organizer and editor of the book, Whitebark Pine Communities: Ecology and Restoration, published by Island Press in 2001, which has grown in significance with the status review of whitebark pine under the Endangered Species Act by the U.S. Fish and Wildlife Service. Dr. Tomback's current research involves studies of the ecological role of whitebark pine at treeline, the effects of whitebark pine mortality on seed dispersal by Clark's nutcracker in the central and northern Rocky Mountains, and the recovery of subalpine forests after the 1988 Yellowstone Fires. In 2001, Tomback and colleagues started the Whitebark Pine Ecosystem Foundation (WPEF) http://www.whitebarkfound.org, a non-profit organization based in Missoula, Montana. The WPEF is dedicated to the restoration of whitebark pine ecosystems through education, information exchange, and outreach. She served as volunteer Director of this organization for 16 years and now as Outreach and Policy Coordinator.

Preface to "Biodiversity and Conservation in Forests"

Global forest communities cover only about 30% of land areas, but they provide important ecosystem services, such as watershed protection, carbon sequestration, and oxygen production, as well as renewable forest products for human subsistence and markets. Forests also support the majority of the world's terrestrial biodiversity. Although land conversion for agriculture and pastureland has historically resulted in fragmentation and declining forested areas, forests worldwide are now experiencing change at an unprecedented rate due to various anthropogenic activities and growing human populations. Global warming trends are altering snowpack and hydrology, fostering outbreaks of native forest pests, and accelerating the loss of older tree age classes. Modeling suggests that future fire regimes in temperate regions will have shorter return intervals, with more severe wildfires. In addition, a by-product of trade and travel globalization has been the accelerated transport of plants and animals, and plant and animal diseases, around the world. Exotic species have altered community composition, especially where foundational tree species are affected. Every forest community worldwide is challenged by some of these problems.

This collection of papers from a special issue on Biodiversity and Conservation in Forests from *Forests* http://www.mdpi.com/journal/forests/special-issues/biodiversity_conservation, broadly illustrates the unique biodiversity supported by forest communities, the range of threats to forest health, how forest communities are rapidly changing, and management and conservation approaches to preserving forest biodiversity. The ten papers (and one erratum note) contributed to this issue, published between May 2015 and April 2016, comprise a diverse geographic representation of conservation and management issues. They include studies from the Peruvian Amazon, Great Barrier Reef and Southwestern Australia, Central Mexico, and Northwestern Yunnan Province in China; from several regions of the United States, including the Northwest, Northeast, Southeast, and from the Rocky Mountains of the United States and Canada.

The contributions, briefly described here, fall into four categories. Under forest biodiversity, there are two papers that present new information on species richness and distribution within two very different forest community types: (1) Yang et al. (2016) compared vegetation zones, plant species richness, species turn-over rates, and environmental influences between the east and west slope of Baima Snow Mountain in the Three Parallel Rivers Region of northwestern Yunnan Province, China, recognized as a UNESCO World Heritage Site for its outstanding biodiversity. The communities on both slopes of Baima Snow Mountain are primarily comprised of coniferous, broad-leaved, or mixed forest types. (2) Small islands and coral cays in the Indian and Pacific oceans support forests dominated by Pisonia (*Pisonia grandis* R.BR.) in different stages of successional development. Rogers et al. (2015) asked whether lichen biodiversity in these communities correlates with understory plant biodiversity and may indicate forest condition in general in Pisonia forests. They examined lichen biodiversity in Capricornia Cays National Park on Heron Island in the Great Barrier Reef to explore these relationships.

In the category disturbance and fragmentation, three papers illustrate the challenges to maintaining healthy and intact forest communities in the face of population growth, land-conversion, economic development, and a changing climate. (1) Lopez-Upton et al. (2015) examined population extinction risk for Douglas-fir (*Pseudotsuga menziesii* (Mirb.) in central Mexico. In this region, Douglas-fir occurs in small, isolated populations, many threatened by anthropogenic activities including over-grazing, land-use changes, and illegal timber harvest, leading in some cases to little regeneration and reduced genetic diversity. (2) The Peruvian

Amazon has experienced large-scale conversion of forest communities to agricultural communities, as well as other anthropogenic disturbances. Perry et al. (2016) suggest that agroforests, which may include a mix of trees and understory crop plants, may help support local biodiversity, and further proposed that insect biodiversity, which is easily sampled, may be a good bio-indicator of habitat integrity. Working in the Ucayali River region near the city of Pucallpa, they compared insect species richness and taxonomic composition among five community types, ranging from natural forests to agroforests to cropland and degraded grasslands. (3) Drought-stress has been associated with damaging outbreaks of woodboring beetles in many forest communities worldwide. More frequent drought conditions and episodic heat waves are predicted to occur with climate change. Seaton et al. (2015) investigate the role of severe drought in outbreaks of the Eucalyptus longhorned borer *Phoracantha semipunctata* in Australian Mediterranean forests dominated by *Eucalyptus marginata* Donn ex Smith and *Corymbia calophylla* (Lindl.) K.D. Hill & L.A.S. Johnson.

The category impacts of invasive exotic species highlights an increasingly dire global problem—the impact of aggressively spreading exotic plants, insects, and pathogens on forest integrity and health. Four papers in this special collection provide instructive examples. (1) Japanese honeysuckle (*Lonicera japonica* Thunb.), an attractive climbing vine, was introduced from China and Japan as an ornamental plant in the early 19th century. It escaped cultivation and is now naturalized in 45 U.S. states and the province of Ontario, Canada. The plant is an economic and ecological threat to natural forests and pine plantations in the southeastern U.S. Wang et al. (2015) documented the rapid spread of Japanese honeysuckle through forestlands in eastern Texas and determined the factors that increase the likelihood of its occurrence. (2) Eastern hemlock (*Tsuga canadensis* (L.) Carr.), a widely-distributed foundation species in the eastern U.S. and Canada, forms ecological communities different in structure and function from neighboring deciduous tree communities. Its dense canopies reduce light penetration and alter nutrient cycling. Eastern hemlock also has multiple influences on riparian and stream communities through the slow decomposition of needles and coarse woody debris. Eastern hemlocks, however, are threatened by the exotic hemlock woolly adelgid (*Adelgis tsugae*), which kills trees by defoliation, resulting in replacement of hemlock by deciduous tree species. In this special issue, we have two papers describing the uniqueness of eastern hemlock communities and the implications of their widespread loss. Working on the Cumberland Plateau in eastern Kentucky, Adkins and Rieske (2015) documented major differences in benthic invertebrate communities in streamside areas dominated by eastern hemlock compared to those areas dominated by deciduous forest trees. (3) Eastern hemlock forests also differ from deciduous forests in understory (herbaceous layer) plant composition and diversity, but the differences vary between the northeastern and southeastern U.S. In the Northeast, the Harvard Forest Hemlock Removal Experiment, Petersham, Massachussets, was established in 2003 to examine potential community-level changes after hemlock loss from hemlock woolly adelgid infestation. Ellison et al. (2015) report on major differences by 2014 among the treatment understory communities, comparing herbs, shrubs, ferns, and grasses and sedges. (4) Whitebark pine (*Pinus albicaulis* Engelm.) ranges throughout the high mountains of the western United States and Canada. As a foundation species, it influences community development and structure through tolerance of poor soils and harsh aspects and effective seed dispersal by Clark's nutcracker (*Nucifraga columbiana*). Its large seeds serve as important wildlife food. Whitebark pine is rapidly declining across its range, and a candidate for listing under the U.S. Endangered Species Act. The primary threat to whitebark pine is white pine blister rust, a disease of five-needle white pines caused by the exotic fungal pathogen *Cronartium ribicola*. Tomback et al. (2016) review whitebark pine's importance in treeline communities in the Rocky Mountains, where it supports snow retention and regulates downstream flows. They document invasion of these communities by white pine blister rust and suggest management actions that could help mitigate the decline.

The final category of papers is ecosystem management. DellaSala et al. (2015) provide a detailed evaluation of a major ecosystem management milestone, the 1994 Northwest Forest Plan, which provided directives for managing federal lands from northern California through Washington State. The multi-agency collaborative plan was developed after "decades of conflict" over the logging of old growth and older forest communities, rapid road expansion, and declining water quality, which resulted in the federal listing of the northern spotted owl (*Strix occidentalis caurina*), the marbled murrelet (*Brachyramphus marmoratus*), and numerous salmonid (*Oncorhynchus spp.*) populations. Providing a 20-year retrospective, DellaSala et al. (2015) address the achievements and limitations of the plan in relation to its original goals, and provide suggestions for future directions, particularly in the light of new scientific information regarding species management and the impact of climate change and other anthropogenic stressors.

The papers in this collection are data-rich and well-suited for stimulating discussion in seminar-style courses and other advanced courses in forest management, ecology, sustainability, and conservation biology. As forest researchers and managers, we have the profound responsibility to inspire and train new generations to take the lead in preserving global forest diversity in the face of growing anthropogenic challenges.

Diana F. Tomback
Special Issue Editor

forests

MDPI

Article

Plant Diversity along the Eastern and Western Slopes of Baima Snow Mountain, China

Yang Yang [1], Zehao Shen [1,*], Jie Han [1] and Ciren Zhongyong [2]

[1] The MOE Key laboratory of Earth Surface Processes; Department of Ecology, College of Urban and Environmental Sciences, Peking University, Beijing 100871, China; lifeyang@pku.edu.cn (Y.Y.); hahj1232008@163.com (J.H.)
[2] Baima Mountain National Reserve, Deqin, Yunnan 674500, China; 295358293@163.com
* Correspondence: shzh@pku.edu.cn; Tel.: +86-10-6275-1179; Fax: +86-10-6275-1179

Academic Editor: Diana F. Tomback
Received: 23 September 2015; Accepted: 18 April 2016; Published: 22 April 2016

Abstract: Species richness and turnover rates differed between the western and eastern aspects of Baima Snow Mountain: maximum species richness (94 species in a transect of 1000 m^2) was recorded at 2800 m on the western aspect and at 3400 m on the eastern aspect (126 species), which also recorded a much higher value of gamma diversity (501 species) than the western aspect (300 species). The turnover rates were the highest in the transition zones between different vegetation types, whereas species-area curves showed larger within-transect beta diversity at middle elevations. The effect of elevation on alpha diversity was due mainly to the differences in seasonal temperature and moisture, and these environmental factors mattered more than spatial distances to the turnover rates along the elevation gradient, although the impact of the environmental factors differed with the growth form (herb, shrubs or trees) of the species. The differences in the patterns of plant biodiversity between the two aspects helped to assess several hypotheses that seek to explain such patterns, to highlight the impacts of contemporary climate and historical and regional factors and to plan biological conservation and forest management in this region more scientifically.

Keywords: elevational gradient; mountain aspect; plant species diversity; comparative study; Baima Snow Mountain; Three Parallel River region; northwest Yunnan

1. Introduction

Mountains are hotspots of biodiversity at global and regional scales [1,2]. The pattern of species richness along the slopes of mountains and the factors that determine such patterns have long been a topic of interest to ecologists and biogeographers [3]. Given the growing variability of climate and changes in land use at present, together with the unprecedented rate, scale and reach of human intervention in mountain habitats, the effectiveness of biodiversity conservation strategies, particularly in the mountains, is becoming a matter for concern [4–6].

The patterns of species diversity along elevation gradients have been studied for decades, and the results of these studies are well summarized [7,8], revealing the major determinants underlying the patterns of diversity, namely environmental filters [9,10], regional processes and evolutionary history [11–13], biological interactions [14] and spatial factors, such as area, dispersal limits and boundary constraints imposed by the elevation range of the mountains [15,16]. However, the collinearity of effects among the different hypotheses has been a critical obstacle to separating the distinct contribution of each of these possible mechanisms [8], a task made even more complex by the difference in the scale of their effects.

Baima Snow Mountain (BSM) is the central section of the eastern-most of three parallel mountain ranges in the Three Parallel Rivers Region (TPRR), northwestern Yunnan Province, China.

By examining plant diversity along the elevation gradient on the eastern and western aspects of this mountain, the present study aimed to explore the effects of elevation and aspect on plant species diversity and to separate the contribution of each of the different mechanisms that determine the structure of vegetation and plant diversity. Specifically, we addressed the following questions: (1) How do vegetation zones and plant species diversity along the elevation gradient on the western aspect of the BSM differ from those on its eastern aspect? (2) What factors affect the local species richness and species turnover along the elevation gradient? (3) How do contemporary environment (current meteorological data and topographic features), historical and regional environmental factors and boundary constraints imposed by the mountain contribute to spatial patterns of plant diversity?

2. Materials and Methods

2.1. Study Area

Field investigations of vegetation were conducted in the northern BSM. The mountain is spread over 2816.4 km², extending chiefly along the south–north axis; its summit rises 5429 m above the mean sea level; the base of its western aspect is at 1815 m, near the Lancang River; and the base of its eastern aspect is at 1950 m, near the Jinsha River (Figure 1). The regional climate is characterized by marked seasonal changes in both temperature and precipitation, received mostly in summer from the south–west monsoon from the Indian Ocean. The interaction between the south–west monsoon and the steep mountain slopes creates a prominent vertical zonation of climate and vegetation, from a dry–warm climate and shrub-dominated vegetation at the base (elevation less than 2600 m) through a cooler and more humid climate and tree-dominated vegetation at middle elevations (2600–4300 m), to the alpine climate dominated by shrubs and grasses at higher elevations (4300 m) [17]. The BSM is divided between two autonomous counties, Deqin and Weixi, in Yunnan province, and was designated as a national natural reserve in 1988 owing to the generally well-protected vegetation and rich biodiversity. Anthropogenic disturbance in the forms of logging and farming is mostly limited to a distance of 500 m from the main road extending into the natural reserves on both eastern and western aspects of the BSM. However, the rich wildlife of the region indicates a harmonious coexistence between the traditional communities and nature. For example, the BSM is the only remaining habitat or species range of the Yunnan snub-nosed monkey (*Rhinopithecusbieti*) [18], a flagship species for biodiversity conservation in Yunnan.

2.2. Meteorological Data

Meteorological data for three years (October 1981–December 1984) were obtained from seven temporary weather stations established across the east–west axis of the BSM (Figure 1, Table 1) by the Interdisciplinary Research Team for Qinghai-Tibet Plateau, the Chinese Academy of Sciences. Three stations were set up on the western aspect, one at the saddle point and the other three on the eastern aspect. Temperature and precipitation data were recorded hourly, and the monthly mean temperature and precipitation values for these sites were published [19]. These are the only data available to compare the eastern and the western aspects of the BSM along its elevation gradient.

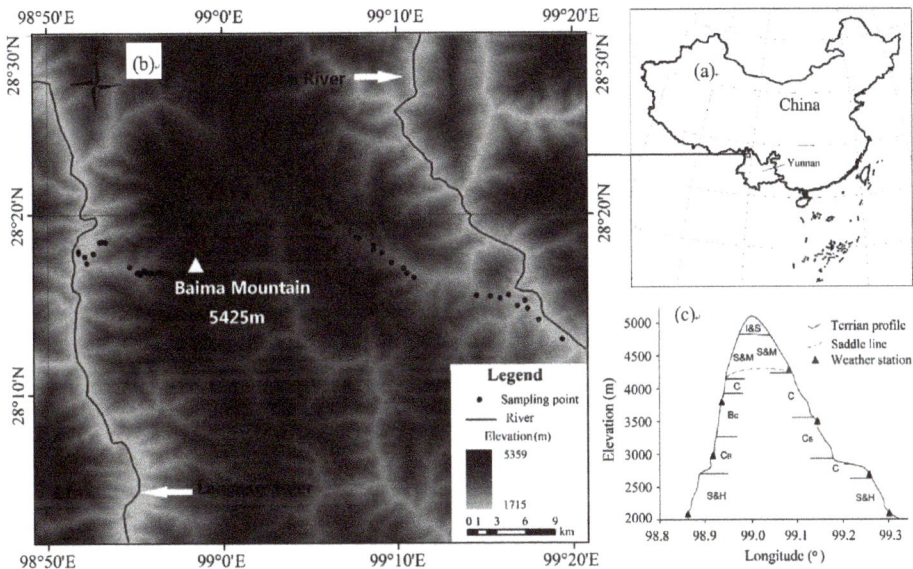

Figure 1. The study area (**a**); distribution of field-sampling sites (**b**); weather stations and the vertical vegetation zones on eastern and western slopes of Baima Snow Mountain (**c**). S&H: shrubs and herbs zone; C: coniferous forest zone; C$_B$: mixed coniferous and broad-leaved forest zone dominated by conifers; B$_C$: mixed broad-leaved-coniferous forest zone dominated by broad-leaved species (mostly evergreen *Quercus*); S&M: alpine shrubs and grasses (meadows) zone; I&S: ice and snow zone.

Table 1. Meteorological data from seven stations on Baima Snow Mountain: 1987–1988.

Parameter	Western Aspect			Saddle		Eastern Aspect	
Station	No.1	No.2	No.3	No.4	No.5	No.6	No.7
Elevation, m	2080	2747	3485	4292	3760	2988	2025
Mean annual temperature, °C	14.74	10.83	5.24	−1.14	2.97	9.65	16.57
Annual precipitation, mm	425.0	410.6	513.8	807.1	946.1	532.8	285.6

We plotted the elevation of the weather stations (black triangles in Figure 1c), the monthly mean temperature (MMT) and the monthly mean precipitation (MMP) at each station for twelve months to calculate the slopes of MMT and MMP along the elevation gradient on the eastern and western aspects and then interpolated the values of MMT and MMP for every 100-m increase in elevation from 2000 m to 4500 m along both the aspects of the BSM. Ten additional bioclimatic indexes were calculated based on the MMT and MMP (Table 2). The patterns of seasonal changes in temperature were similar for both aspects, but those of seasonal changes in precipitation were distinct. The elevation at which precipitation was maximum on the western aspect could not be determined precisely (because it kept increasing with elevation), whereas the elevation could be determined on the eastern aspect eight months of the year (except May, June, July and August). The bottoms of the slopes towards the Lancang River (western aspect) and the Jinsha River (eastern aspect) were both characterized by a hot–dry climate (especially on the eastern aspect), and precipitation increased much faster at higher elevations. Net primary productivity (NPP) was calculated using the data and the formula given in Table 2.

Table 2. Climatic variables used in the data analysis.

Factors	Index	Algorithm	Reference
Energy	Tmin	Monthly mean temperature in the coldest month (January)	[20]
	Tmax	Monthly mean temperature in the warmest month (July)	[20]
	MTWQ	Mean temperature in the wettest quarter: June, July and August	[20]
	MTCQ	Mean temperature of December, January and February	[20]
	MAT	Mean annual temperature	
	PET	Potential evapotranspiration: $58.93 \times$ ABT [†]	[21]
Moisture	PWQ	Precipitation in the wettest quarter: June, July and August	[21]
	PDQ	Precipitation in the driest quarter: December, January and February	[21]
	Pmin	Minimum monthly precipitation (in January)	
	Pmax	Maximum monthly precipitation (in August)	
	AP	Annual precipitation	
	AET	Actual evapotranspiration: $P/(0.9 + (P/L)^2)^{1/2}$, $L = 300 + 25T + 0.05T^3$	[22]
	MI	Moisture index: PET/AP	[21]
	WD	Water deficiency: $PET - AET$	[21]
	DI	Drought index: AET/P	[21]
Productivity	NPP	Net primary productivity: $\min\{NPP_{MAP}, NPP_{MAT}\}$ $NPP_{MAP} = 0.005212(MAP^{1.12363})/e^{0.000459532(MAP)}$ $NPP_{MAT} = 17.6243/(1 + e^{(1.3496 - 0.071514(MAT))})$	[23]
Climate seasonality	ART	Annual range of temperature: $T7 - T1$	[20]
	TSN	Temperature seasonality: SD (monthly mean temperature) $\times$ 100	[20]
	PSN	Precipitation seasonality: CV (monthly mean precipitation)	[20]

[†] ABT: annual Biotemperature (°C). ABT $= \sum t_i/12$. t_i is mean month temperature that larger than 0 °C, when $t_i > 30$ °C, it is assigned to 30 °C.

2.3. Field Sampling

Vegetation sampling was conducted in 2012 and 2013. Starting at 2000 m, we established a transect every 100 m along the elevation gradient, up to 4300 m on the western aspect (a total of 24 transects) and up to 4400 m on the eastern aspect (25 transects). Each transect was 10 m wide and 100 m long, with the shorter dimension along the gradient and the longer dimension across the gradient. For easy access, the sites along the road across the mountain were chosen for representation and at least 500 m away from the road to reduce the influence of human intervention. Each transect was established on a slope that was uniform in topography. Because the weather stations were also along the road, the three-year data form a spatially-representative sample of the climate of the transects.

Each transect was divided equally into 10 plots (10 m $\times$ 10 m). In each plot, the height and the girth (circumference) at breast height (1.3 m) of all of the trees were measured and the species recorded. For each species of shrub and herb, all plants within a plot were counted as abundance, and the percentage of coverage was estimated visually. The geographic coordinates and elevation of each transect were recorded using a GPS, and the gradient of the slope and its direction were measured with a compass.

2.4. Statistical Analysis

2.4.1. Estimate of α and β Diversity

Species richness of all vascular plants within a transect was counted as local species richness (or α diversity), and that of trees, shrubs and herbs (including ferns) was also counted separately. β diversity was estimated using the Simpson dissimilarity index (β_{sim}) [24] for species turnover rate between neighboring transects along the elevation gradient on the western and the eastern aspects separately. β_{sim} was calculated between adjacent transects, for all species within a transect, and for trees, shrubs and herbs separately, and only the presence or absence of each species in the transect was used in the calculations with the following formula:

$$\beta_{sim} = \frac{min\,(b,c)}{a + min\,(b,c)} \tag{1}$$

where *a* is the number of species shared between two transects, *b* is the number of species present in transect B, but not in transect C, *c* is the number of species present in transect C, but not in transect B, and min () indicates the smaller of the two values between *b* and *c*. β_{sim} was used because it provides a fairly reliable estimate of species turnover independent of the impact of local species richness [25].

Based on the species composition of ten 10 m × 10 m plots in each transect, a species-area curve was used to describe the changes in species diversity with increasingly larger sampling areas [26]. Species-area curves for pairs of transects at the same elevation, but on different aspects (western or eastern) of the BSM were compared to find the differences in complexity of the community structures between the two aspects. Species richness of all possible combinations of a certain number *i* (*i* = 1, 2, 3, . . . ,10) of unit plots was calculated, and the means and standard deviations of species richness were used to draw the species-area curve for each transect, with the area increasing from 100 m^2 to 1000 m^2.

2.4.2. Principal Component Analysis (PCA) of Climatic Variables

Because multiple bioclimatic variables are collinear, PCA was applied to extract principal bioclimatic information with focused indices. PCA is useful in reducing the number of dimensions of explanatory variables with acceptable information loss under most conditions [27]. All 18 climatic variables were classified into three types, namely energy, moisture and climatic seasonality. Before PCA, all indexes were normalized to have a mean of zero and a standard deviation of 1. The first principal component of three energy indexes (Energy.pc1) loaded 99.5% of the variation in energy; the first two principal components of seven moisture indexes (Moist.pc1, Moist.pc2) accounted for 97.4% of the variation in moisture, indicating the importance of rainfall during the growing season and of winter precipitation, respectively; and the first two principal components of six indices of climatic seasonality (Seasonal.pc1, Seasonal.pc2), indicating seasonal changes in temperature and precipitation, respectively, accounted for 99.9% of the variation (Table S1).

2.4.3. Environmental and Spatial Interpretations of Diversity Patterns

A hierarchical variation partitioning (HVP) algorithm based on the generalized linear model (GLM) was used for examining the independent influence of environmental variables on species richness: we applied the HVP algorithm to species richness of all species and to tree, shrub or herb species separately. The algorithm creates a GLM of all possible combinations of the explanatory variables, uses Akaike's information criteria (AIC) to select the optimal model and estimates the independent contribution of each of the considered variables to the selected model [28–30]. We used the HVP algorithm for the variables included in the final model of the environmental interpretation of α diversity patterns.

The Mantel test and the partial Mantel test are the two commonly-used methods of testing the association between two matrices, the entries of which are distances or similarities [31]. As an extension of the partial Mantel test, a multiple regression model (MRM) involves multiple regressions of a response matrix on a given number of explanatory matrices, which contain spatial distances or environmental similarities between all pairs of sampled units [32]. The model is useful in decomposing the collinear effects of space and environmental factors on the spatial patterns of β diversity, and we applied the MRM to the species composition of each transect considering all of the plant forms together and also the composition of tree, shrub and herb species separately. Euclidian distance was used for calculating the spatial distance and environmental dissimilarity between two transects, and latitude, longitude and elevation were used for calculating the spatial distance matrix. The dissimilarity matrix for all transects (24 on the western aspect and 25 on the eastern aspect) was calculated separately for the indexes of energy, moisture, climatic seasonality, net primary productivity (NPP) and a comprehensive environmental index. To validate the results of MRM, we also applied the Mantel test and the partial Mantel test to estimate the impact of spatial distance and of environmental differences in energy, moisture, climatic seasonality and productivity on the rate of species turnover between transects.

3. Results

3.1. Vegetation Zones and Overall Species Composition along the Elevation Gradient

The vegetation zones showed distinctly different patterns between the western and the eastern aspects of the BSM (Figure 1), although the overall patterns of plant diversity on both sides were broadly similar: shrubs and herbs in the dry-warm climate at the bottom, trees at middle altitudes and shrubs and grasses in the alpine climate at higher elevations. The upper limit of the lower zone on both aspects was 2700–2900 m, above which the vegetation transitioned to the forest zone through a broad tree line ecotone. The structure of the forest zone along the elevation gradient differed between the two aspects: the western aspect showed the "C_B-B_C-C" pattern, which consisted of mixed coniferous and evergreen broad-leaved forests that were dominated by *Pinus* species (C_B, 2800–3300 m); mixed evergreen broad-leaved and coniferous forests that were dominated by *Quercus* species (B_C, 3300–3900 m); and a coniferous forest composed of *Abies* and *Larix* species (C, 3900–4200 m); the eastern aspect displayed a distinct "C-C_B-C" pattrn, *i.e.*, C dominated by *Pinus* species at 2700–3000 m, C_B at 3000–3600 m and C composed of *Abies* and *Larix* species at 3600–4300 m. The two aspects also showed distinct community structures across the forest zone (Figure 2). On the western aspect, the sum of the basal area of all trees (SBAT) in a transect (as a proxy for forest biomass) increased with elevation, reaching its maximum value near the upper limit of the subalpine conifer forests. On the eastern aspect, however, the SBAT showed two peaks, one at 3400 m and one at 4200 m. The contribution of coniferous trees to the SBAT showed a U-shaped pattern on the western aspect, with conifers dominating at lower elevations (2900–3400 m), as well as at higher elevations (4000–4200 m) and broad-leaved trees dominating at the middle elevations (3500–3900 m). On the eastern aspect, however, conifers dominated nearly all of the forest zones except at middle elevations (3200–3500 m), where the communities were equally rich in both conifers and broad-leaved trees (Figure 3, Figure S1).

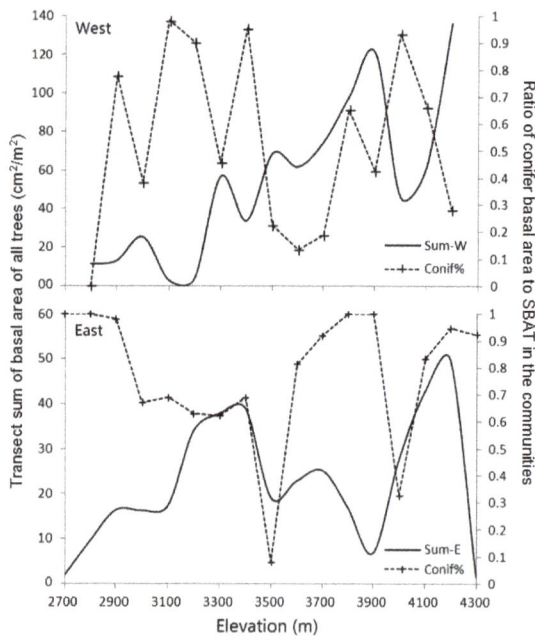

Figure 2. Elevation-related patterns of the transects: sum of basal area of all trees (SBAT, solid line) and the ratio of conifer basal area to SBAT in the communities (broken line with crosses) on western and eastern aspects of Baima Snow Mountain.

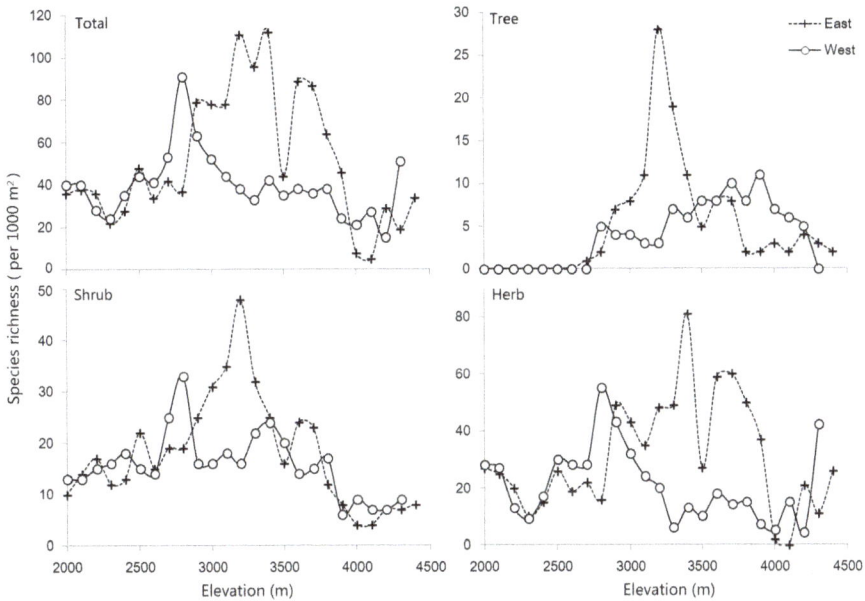

Figure 3. Elevation-related patterns of species richness of all plants and of trees, shrubs and herbs in the transects on western and eastern aspects of Baima Snow Mountain.

Species diversity on the eastern aspect of the BSM was higher than that on the western aspect. The eastern aspect harbored a total of 501 species of vascular plants (304 herbs, 155 shrubs and 42 tree species), representing 250 genera and 94 families; the corresponding figures for the western aspect were 300 (189, 92, and 19), comprising 194 genera and 82 families.

3.2. Species Richness and Species Turnover within Communities

Species richness at the transect level varied with elevation, but showed different patterns on the western and the eastern aspects (Figure 3). In general, species richness of vascular plants over sampling areas of 1000 m^2 each described a hump-shaped curve on both the aspects, reaching its lowest value at the upper limit of the forest zone (about 4000 m), whereas species richness of shrubs and herbs increased with elevation. Total species richness and that of shrubs and herbs peaked at 2800 m on the western aspect, in the ecotone between the lower dry shrubs-and-grasses zone and the mid-altitude forest zone. On the eastern aspect, species richness of all of the three growth forms—herbs, shrubs, and trees—peaked at 3200–3400 m. Species richness on both aspects was very similar at lower elevations (in the dry valleys below 2700 m) and also at higher elevations (the alpine shrubs-and-grasses zone above 4000 m). The greatest contrast in species richness between the two aspects occurred in the forest zone (2800–3900 m), a difference that was also evident in the structure of vegetation mentioned earlier (Figure 1).

The species-area curves of transects at the same elevation on the western and the eastern aspects were also hump shaped (Figure 4), with maximum species richness in all ten unit plots occurring at middle elevations, roughly 2800–3700 m. The two curves came close at elevations between 2000 m and 2700 m, indicating similar α diversity and within-transect β diversity between the communities on the western and the eastern aspects. At 2800 m, the transect on the western aspect had a much larger α diversity and within-transect β diversity (indicated by the increase of species richness along with the increasing area) than that on the eastern aspect. However, the relationship was reversed in most of the forest zone. From 2900–3900 m, transects on the eastern aspect had larger α diversity and within-transect β diversity than those on the western aspect. At 4000–4300 m, in the subalpine conifer forests on both aspects, the differences between the species-area curves gradually decreased, and α diversity and within-transect β diversity tended to be greater on the western aspect more frequently.

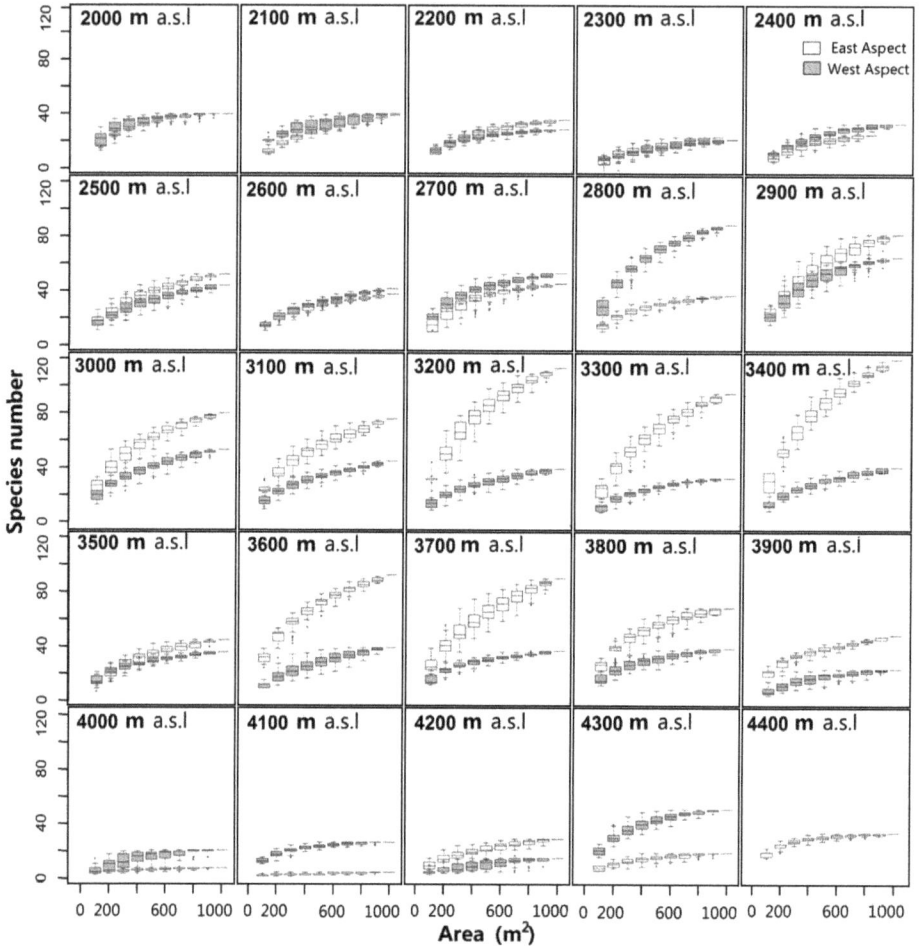

Figure 4. Contrast of species-area curves between pairs of transects along the elevation gradient on eastern and western aspects of Baima Snow Mountain. The horizontal line in each boxplot indicates the median species richness in a specific sampling area at different elevations and aspects and the error bar indicates 2.5 and 97.5 quantiles.

3.3. Vegetation Zones and Species Turnover

The average species turnover rate along the elevation gradient on the eastern aspect (mean ± SD = 0.53 ± 0.37) was greater (t = −3.51, $p < 0.001$) than that on the western aspect (0.36 ± 0.19) (Figure 5), and this difference was significant for shrubs (0.50 ± 0.25 *vs.* 0.33 ± 0.15; t = −2.88, $p = 0.006$) and herbs (0.55 ± 0.21 *vs.* 0.39 ± 0.17; t = −3.013, $p = 0.004$), but not for trees (0.25 ± 0.23 *vs.* 0.22 ± 0.21; t = −0.89, $p = 0.38$). On both aspects, the species turnover rate was consistently the highest among herbs (0.47 ± 0.21) and the lowest among trees (0.23 ± 0.22), and the differences among all three growth forms were significant.

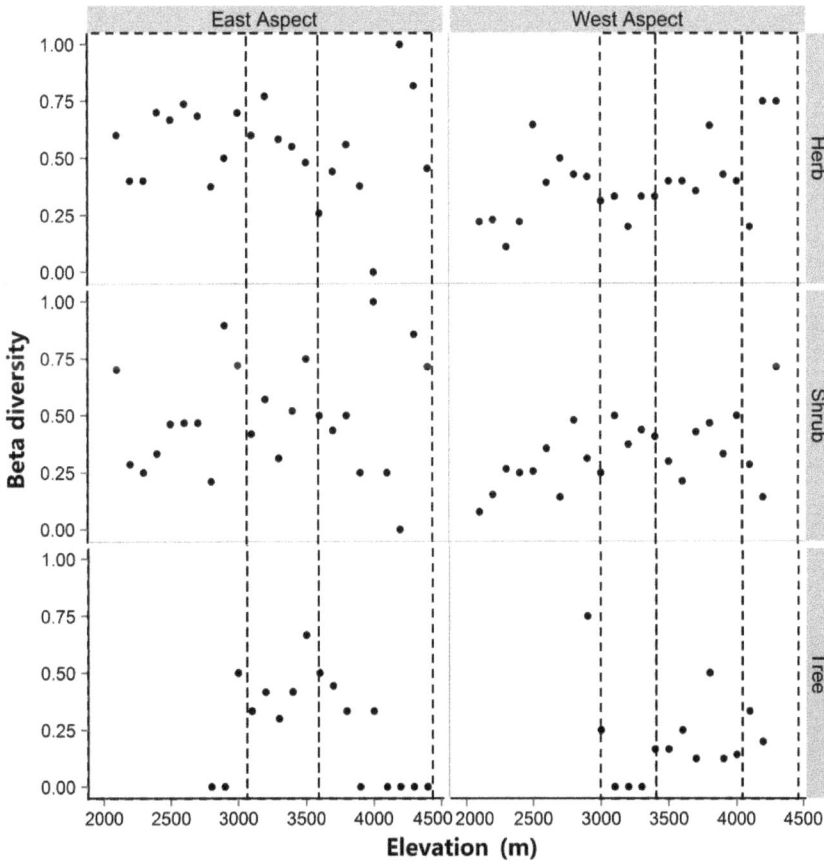

Figure 5. Species turnover rates of all plant species and of trees, shrubs and herbs along an elevation gradient on eastern and western aspects of Baima Snow Mountain.

No consistent pattern of species turnover rate (or across-transect β diversity) was found along the elevation gradient, either between the different growth forms or between the two aspects. As expected, a transition between two neighboring vegetation zones or sub-zones generally corresponded to an increase in the species turnover rate. On the western aspect, maximum rates of species turnover occurred at 2700–2800 m, 3800 m and 4200–4300 m, corresponding to the transitions between the dry valley shrubs-and-herbs zone and the forest zone, between the mixed-forests zone (broad-leaved and coniferous) and the conifer-forests zone and between the forest zone and the alpine shrubs-and-grasses zone. On the eastern aspect, maximum rates of species turnover occurred at 2900–3000 m, 3500 m and 4300 m, elevations that corresponded to the similar transitional zones on the western aspect.

3.4. Environmental Interpretation of Alpha Diversity Patterns

With the HVP algorithm, most of the variation (78.8%–95.8%) in species richness, whether for all of the growth forms together or for each of the growth forms separately, along the elevation gradient can be accounted for by the environmental factors (Table 3). In general, the proportions of species richness patterns on both of the aspects accounted for by the models were comparable, although the proportion of variation in species richness left unexplained was greater in the case of herbs than that of shrubs or trees. The environmental influence on the variation in species richness of the three growth forms was more consistent on the eastern aspect than that on the western aspect and was primarily due to changes in seasonal temperature (annual range of temperature (ART); and temperature seasonality (TSN)), although the influences of slope and that of precipitation during the growth season (precipitation in the wettest quarter (PWQ); maximum monthly precipitation (Pmax); and actual evapotranspiration (AET)) were also important. The species richness of trees was primarily related to energy (potential evapotranspiration (PET); mean annual temperature (MAT); and minimum monthly temperature (Tmin)); of shrubs, to moisture (minimum monthly precipitation (Pmin); Pmax; and PWQ); and of herbs, to temperature (ART, TSN). The local topographic features had little impact on species richness patterns on the eastern aspect.

Table 3. Hierarchical variation partitioning of elevation-related species richness patterns of herb, shrub and tree species and of all species on the western and the eastern aspects of Baima Snow Mountain. Refer to Table 2 for abbreviation and definitions.

Variable	Western Aspect				Eastern Aspect			
	Herbs	Shrubs	Trees	All Forms	Herbs	Shrubs	Trees	All Forms
MAT		0.080	0.174 *		0.056	0.072	0.082	0.089
Tmin		0.087	0.174 *	0.107				
PET		0.118	0.281 *	0.107	0.051		0.166 *	0.078
Pmin		0.169 *			0.100			
Pmax		0.122 *		0.211 *			0.075	0.070
PDQ	0.306 *				0.091			0.107
PWQ		0.122 *		0.201 *	0.048	0.073	0.081	0.070
WD		0.073		0.114 *				
AET			0.167 *		0.050	0.123	0.064	0.072
NPP	0.068				0.159 *			
ART	0.223 *	0.096			0.110	0.238 *	0.178 *	
TSN	0.200 *			0.169 *		0.257 *	0.171 *	0.221 *
PSN			0.072					
Exposure	0.061				0.042	0.028		0.027
Slope					0.174 *	0.077		0.171 *
Interpreted variation (%)	85.7	88.8	92.7	90.2	83.6	95.2	95.8	92.5

* Significant at 95% by the Z test.

3.5. Climatic Interpretation of Beta Diversity Patterns

The results from MRM indicated that almost all climatic variables and the spatial distance significantly influenced the dissimilarity in species composition between the transect communities and that the impact of climatic variables was greater than that of spatial distance (Table 4). The most prominent climatic variable on both aspects was moisture in the case of changes in species composition of herbs and shrubs and energy in the case of trees: the results accounted for 62.8%–64.5% of the community dissimilarity on the western aspect, but only 46.5%–47.2% of that on the eastern aspect.

The partial Mantel test (Table S3) also indicated a generally larger impact of climatic variables than that of spatial distance on community dissimilarity. Furthermore, the impact of geographic distance on the dissimilarity was not significant for any of the growth forms in the case of the eastern aspect, whereas the impact of climatic variables was generally significant. On the western aspect, the impact of climatic variables was significant and so was that of geographic distance in the case of shrubs and herbs, although not in the case of trees. Specifically, the most prominent climatic factor accounting for the dissimilarity in shrubs and herbs was NPP (*i.e.*, the water–energy balance) on the eastern aspect and moisture on the western aspect.

Table 4. Regression coefficients and regression R-squared values: multiple regression on distance matrices of beta diversity on the eastern and the western aspects of Baima Snow Mountain. D_e and D_g: Euclidian distance between two transects calculated for environmental variables or geographic coordinates (including elevation), respectively.

Distance	Variables	Eastern Aspect			Western Aspect		
		Herb	Shrub	Tree	Herb	Shrub	Tree
D_e	Energy	0.156	0.335	0.995 ***	0.180	0.035	1.269 *
	Moisture	0.385 **	0.525 **	0.289	0.513 **	0.618 **	1.133 *
	Seasonality	0.076	0.052	0.035	0.211 *	0.321 **	0.606
	Productivity	0.172 *	0.217 **	0.035	0.204 **	0.234 **	0.224
D_g		0.359 *	0.477 **	0.682 ***	0.230	0.354 **	0.410 *
Total	R^2	47.2% ***	47.1% ***	46.5% ***	62.8% ***	64.6% ***	63.8% ***

*, $p < 0.05$; **, $p < 0.01$; ***, $p < 0.001$.

4. Discussion

4.1. Interpretation of Patterns

For the BSM, the major source of precipitation is the Indian monsoon coming from the west [17], which normally arrives in May; the rainy season lasts until October [19]. Before it reaches the BSM, the monsoon crosses two mountain ranges (namely Mount Gaoligong and Mount Yunling) oriented along the north–south axis in the TPRR and deposits most of the precipitation on their windward slopes. Therefore, the precipitation on the windward slope of the BSM is much lower at the bottom of the valley and increases with elevation. Rainfall is maximum over the mountain ridge on the leeward side (the eastern aspect) at 3800–3900 m and decreases sharply as the elevation decreases [19]. Therefore, the annual mean precipitation at the lower section of the bottom of the valley (below 2500 m) on the eastern aspect of the BSM was less (285.6 mm) than that on the western aspect (425 mm). At middle and upper elevations (2600–3300 m and 3400–4200 m), however, precipitation on the eastern aspect (532.8 and 946.1 mm, respectively) was greater than that on the western aspect (440.3 and 513.8 mm, respectively) [19]. Because the transects on both the aspects were established at similar elevations, the difference in moisture levels due to the monsoon seems to be the major climatic factor influencing vegetation patterns and plant species diversity on the two aspects.

The Hengduan Mountains are a global hotspot of biodiversity [1]. In the southern part of the Hengduan Mountains, Yunnan province harbors half the plant species pool in China and two endemism centers [33], and the TPRR in northwestern Yunnan is one of them. The general pattern of biodiversity along the elevation gradient in this region is described elsewhere [34–38], but other characteristics of the distribution of species richness are not widely known, especially with regard to the impact of the Indian Ocean monsoon, a primary driver of the region's climate. In fact, plant species richness on the eastern aspect is higher than that on the western aspect of the BSM, especially within the forest zone, which also has a wider range on the eastern aspect (2700–4300 m) than on the western aspect (2800–4200 m). The evergreen oak forests in the Hengduan Mountains dominate the dry and sunny habitats [39,40], and the dominance of the *Quercuspanosa* community on the western aspect of the BSM also indicates a habitat drier than the mixed coniferous-deciduous broad-leaved forests, which cover a much wider range of elevations on the eastern aspect (Figure 1).

4.2. Testing Competing Hypotheses

Past investigations of spatial patterns of biodiversity have narrowed the explanations to several mechanisms that govern such patterns, and these mechanisms require further study. The mechanisms include the effects of area, climate, environmental heterogeneity, historical and regional processes, as well as geometric constraints. In the species-area theory, different models predict that species richness increases with sampling area [26]. Climate-related hypotheses suggest that biodiversity is limited by low temperatures, by cumulative warmth or by the water–energy balance indicated by the net primary productivity (summarized by Wang *et al.*, 2009 [41]). The environmental heterogeneity hypothesis predicts a positive correlation between niche diversity and species richness [8]. The historical/regional process hypothesis suggests that local species diversity is limited by the regional species pool, which is mainly regulated by the spatial patterns of climate and geomorphology on a regional scale, as well as by their historical processes, especially in the Quaternary period [8,42,43]. A geometric constraint occurs when species are randomly distributed in a domain with only a single geometric boundary constraint, and the ranges of their distribution collectively generate a hump-shaped species richness pattern along one spatial dimension, showing the middle-domain effect (MDE) [8]. The MDE hypothesis thus predicts a hump-shaped species richness pattern along an elevation gradient irrespective of the impacts of other environmental gradients [44]. Earlier studies have provided support for the influence of climatic factors, historical and regional processes and geometric constraints on elevation-related patterns of biodiversity in the TPRR [34–36], although consensus on the relative contributions of these factors is lacking.

By sampling the same area at 100-m intervals of elevation, sampling similar numbers of transects (24 and 25) and sampling similar elevation ranges (2000–4300 m and 2000–4400 m) on both aspects of the BSM, our transect data controlled for the area effect on species richness. Given the rise of the Tibet-Qinghai Plateau and the simultaneous erosion of the valleys of the Lancang River and the Jinsha River on both sides of the BSM, the glacial and interglacial cycles in the Quaternary period and the migration of plant species on the two aspects in response to these changes can also be reasonably regarded as similar. Therefore, the present study gathered a dataset useful for comparing elevation-related patterns of plant species diversity on the eastern and the western aspects of the BSM and for testing the above-mentioned hypotheses that explain the patterns of diversity (Table 5).

Table 5. Comparison of different hypotheses to explain the elevation-related patterns of plant species diversity in Baima Snow Mountain. α diversity, species richness of transects; β diversity, species turnover rate between pairs of neighboring transects; r diversity, total species recorded in all transects of each (eastern or western) aspect of the BSM; ×, rejected; $\sqrt{}$, supported; $\not{\sqrt{}}$, partially supported (or rejected).

Interpretive Hypothesis	α Diversity	β Diversity	R Diversity	Species-Area Curves
Area	×	×	×	$\not{\sqrt{}}$
Middle-domain effect	$\not{\sqrt{}}$	$\not{\sqrt{}}$	×	×
Historical and regional process	$\sqrt{}$	×	$\sqrt{}$	$\not{\sqrt{}}$
Water-energy balance	$\sqrt{}$	$\sqrt{}$	$\sqrt{}$	$\sqrt{}$
Habitat heterogeneity	$\not{\sqrt{}}$	×	×	$\sqrt{}$

For α diversity of the plant communities, the area hypothesis cannot explain the elevation-related variation in plant species richness, and the MDE hypothesis cannot be completely rejected by the generally hump-shaped pattern of elevation-related species richness. The elevation-related climatic variables (especially precipitation and seasonal changes in climatic variables), although based on a limited monitoring period of three years, correlate well with the α diversity patterns (Table 3 and Table S1). Similar results have also been reported earlier [35,36]. Historical and regional processes can also predict this pattern, because the up-and-down migration of plant species during the glacial and interglacial cycles can also increase the possibility of more species ending up at middle elevations [36]. Habitat heterogeneity, although not specifically measured, cannot explain the huge difference in species richness (3–118) along the elevation gradient, unless it is linked with the variation in climate-related energy.

Neither area nor the MDE model can provide any support to the elevation-related patterns of between-transect species turnover rate (β_{sim}), although changes in climatic variables had a greater impact on that rate than spatial distance did (Table 4). Since elevation-related vegetation zones reflect climatic zones, the match between peak values of species turnover rate and transitions in the type of vegetation showed the effects of an elevation-related climate gradient on beta diversity patterns. It is reasonable to infer that an elevation-related climate gradient cannot be a more prominent cause of elevation-related patterns of beta diversity than habitat heterogeneity within each transect. The effect of spatial distance on beta diversity is linked to the limitations on species dispersal, which can be affected by regional geomorphology [45]. The characteristically steep topography of the BSM was a statistically-significant, but much weaker predictor of species richness than the environmental differences between transects.

All species recorded in all of the transects on the western and the eastern aspects were comparable in terms of their gamma (r) diversity. The differences in biodiversity between the eastern and the western slopes—501 *vs.* 300 species, 250 *vs.* 194 genera and 94 *vs.* 82 families, respectively—are among the major results of the present study. These differences could not be predicted by similar sampling areas, identical elevation boundaries of the studied domain (BSM) and within-habitat heterogeneity: none of the three factors could provide a reasonable prediction. However, the greater precipitation on the eastern aspect (at 2600–4300 m)than on the western aspect most likely contributed more to the productivity of the vegetation and, therefore, acted as a regional explanation of the difference in gamma diversity between the two aspects.

Although species richness increased with the sampling area in all transects, neither area nor geometric constraint can account for the elevation-related differences between the two aspects as seen in the species-area curves. However, the differences in the species pool as a regional factor can account for the consistently higher diversity at all scales on the eastern aspect, and the more complex canopy structure of the vegetation community on the eastern aspect might also increase the heterogeneity of the local habitat, which is critical to herb and shrub diversity.

In summary, the elevation-related patterns and the differences between the two aspects, eastern and western, of the BSM in terms of plant species diversity received prominent support from the influence of contemporary environment (primarily climate) and regional-scale environmental differentiation related to historical process, whereas the impact of area and geometric constraint was found to be somewhat weak. Although local habitat heterogeneity was not measured directly, there was no indication that it may have a major impact on plant species richness pattern on the scale of the entire elevation gradient and between the two aspects.

The large proportion of unexplained variation in the models, especially for beta diversity, indicates uncertainty in the results, and caution is warranted in interpreting them. Owing to the short-term records of climate and data on biodiversity based on only one transect at every 100-m elevational distance along each of the aspects of the mountain, further work is necessary to reveal the general differences between the two aspects of the BSM in terms of elevation-related variation in plant diversity.

4.3. Implications

The TPRR, including the BSM, is listed as a United Nations World Heritage site because of its exceptional biodiversity, which comprises 20% of the total plant species richness and a quarter of the vertebrate species richness of China, and its cultural diversity, which encompasses twelve different ethnic groups [46]. Humans settled this region thousands of years ago, obviously adapted to its natural resources and environment. Since human activities are strictly prohibited in natural reserves, a conflict between the modern approach of biological conservation and the traditional strategy of natural resource management by the local communities was inevitable and sometimes intense, as has been reported from all over the world [47,48].

Therefore, it is particularly important to know more about the distribution of biodiversity within the huge expanse of TPRR and to identify priority areas for biodiversity conservation. Although one of the global centers of plant diversity, especially of gymnosperms [49], the spatial distribution of species in this region, except the patterns related to elevation, has rarely been studied. Our study provides the first community-based assessment and data on the differences between the eastern and the western aspects of the BSM in terms of elevation-related patterns of plant species richness and changes in the species composition. The mixed coniferous and deciduous broad-leaved forests and the evergreen *Quercus* forests are two dominant forest types in this region, typically occupying the more humid or the drier aspects of the TPRR. Understanding the differences in plant diversity in these two forest types is helpful in biodiversity conservation and natural management. For example, the extraordinarily high diversity of conifer species on the eastern aspect at 3200–3300 m corresponds to the maximum plant species richness across the elevation gradient (Figure S1). Since it included ten species of conifers, representing three families and six genera, the local habitat serves as a micro-refugium for a large number of plant species and is probably a biodiversity hotspot within this region. The environmental requirements of these species deserve further study, which will inform the design of sound conservation plans and management of human–nature conflicts.

5. Conclusions

Although elevation-related patterns of species diversity and the underlying mechanisms have been widely discussed, few studies have compared the patterns between different aspects of the same mountain, and the present study proved to be a unique and helpful opportunity to test competing hypotheses that seek to explain such patterns. We were able to separate the distinct roles of different mechanisms that may affect the elevation-related plant diversity patterns in the BSM, to highlight the role of changes in climatic factors along the elevation gradient and to identify the difference in the effect of the monsoon between the eastern and the western aspects of the mountain. The study highlights the differences due to the aspect in patterns of plant biodiversity along the elevation gradient and provides useful insights for planning biodiversity conservation and forest management in this region.

Supplementary Materials: Supplementary materials are available online at http://www.mdpi.com/ 1999-4907/7/4/89.

Acknowledgments: This study was sponsored by National Natural Science Foundation of China (41371190, 31170449, 31321061). We are grateful to Diana F. Tomback and two anonymous reviewers for their help in improving the draft and language.

Author Contributions: Z.S. designed the study. Z.S., Y.Y., H.J. and C.Z. did the experiment. Y.Y. and Z.S. did the analysis. Z.S. and Y.Y. wrote the manuscript.

Conflicts of Interest: The authors declare no conflict of interest.

References

1. Myers, N.; Mittermeier, R.A.; Mittermeier, C.G.; Fonseca, G.A.D.; Kent, J. Biodiversity hotspots for conservation priorities. *Nature* **2000**, *403*, 853–858. [CrossRef] [PubMed]
2. López-Pujol, J.; Zhang, F.M.; Ge, S. Plant biodiversity in China: Richly varied, endangered, and in need of conservation. *Biodivers. Conserv.* **2006**, *15*, 3983–4026. [CrossRef]
3. Sanders, N.J.; Rahbek, C. The patterns and causes of elevational diversity gradients. *Ecography* **2012**, *35*, 1–3. [CrossRef]
4. Wu, R.; Zhang, S.; Yu, D.W.; Zhao, P.; Li, X.; Wang, L.; Yu, Q.; Ma, J.; Chen, A.; Long, Y. Effectiveness of China's nature reserves in representing ecological diversity. *Front. Ecol. Environ.* **2011**, *9*, 383–389. [CrossRef]
5. Sloan, S.; Jenkins, C.N.; Joppa, L.N.; Gaveau, D.L.A.; Laurance, W.F. Remaining natural vegetation in the global biodiversity hotspots. *Biol. Conserv.* **2014**, *177*, 12–24. [CrossRef]
6. Ying, L.; Shen, Z.; Chen, J.; Fang, R.; Chen, X.; Jiang, R. Spatiotemporal patterns of road network and road development priority in Three Parallel Rivers region in Yunnan, China: An evaluation based on modified kernel distance estimate. *Chin. Geogr. Sci.* **2014**, *24*, 39–49. [CrossRef]
7. Rahbek, C. The elevational gradient of species richness: A uniform pattern? *Ecography* **1995**, *18*, 200–205. [CrossRef]
8. McCain, C.M.; Grytnes, J.-A. Elevational gradients in species richness. In *Encyclopedia of Life Sciences*; John Wiley&Sons, Ltd.: Chichester, UK, 2010.

9. Gentili, R.; Armiraglio, S.; Sgorbati, S.; Baroni, C. Geomorphological disturbance affects ecological driving forces and plants turnover along an altitudinal stress gradient on alpine slopes. *Plant. Ecol.* **2013**, *214*, 571–586. [CrossRef]
10. Brown, J.H. Mammals on mountainsides: Elevational patterns of diversity. *Glob. Ecol. Biogeogr.* **2001**, *10*, 101–109. [CrossRef]
11. Hawkins, B.A.; Turner, J.R.G. Energy, water, and broad-scale geographic patterns of species richness. *Ecology* **2003**, *84*, 3105–3117. [CrossRef]
12. Ricklefs, R.E.; Qian, H.; White, P.S. The region effect on mesoscale plant species richness between eastern Asia and eastern North America. *Ecography* **2004**, *27*, 129–136. [CrossRef]
13. Wiens, J.J.; Parra-Olea, G.; Wake, D.B. Phylogenetic history underlies elevational biodiversity patterns in tropical salamanders. *Proc. R. Soc. Lond. B Biol.* **2007**, *274*, 919–928. [CrossRef] [PubMed]
14. Terborgh, J.; Weske, J.S. The role of competition in the distribution of Andean birds. *Ecology* **1975**, *56*, 562–576. [CrossRef]
15. Colwell, R.K.; Hurtt, G.C. Nonbiological gradients in species richness and a spurious Rapoport effect. *Am. Nat.* **1994**, *144*, 570–595. [CrossRef]
16. Lomolino, M.V. Elevation gradients of species-density: Historical and prospective views. *Glob. Ecol. Biogeogr.* **2001**, *10*, 3–13. [CrossRef]
17. Li, H.W.; Zhao, Y. Plant diversity of Baimaxueshan National Natural Reserve. *Guihaia* **2007**, *27*, 71–76.
18. Kirkpatrick, R.C.; Long, Y.C.; Zhong, T.; Xiao, L. Social organization and range use in the Yunnan snub-nosed monkey *Rhinopithecus bieti*. *Int. J. Primatol.* **1998**, *19*, 14–51. [CrossRef]
19. Zhang, Y.G. Several issues concerning vertical climate of the Hengduan Mountains. *Res. Sci.* **1998**, *20*, 12–19.
20. Hijmans, R.J.; Cameron, S.E.; Parra, J.L.; Jones, P.G.; Jarvis, A. Very high resolution interpolated climate surfaces for global land areas. *Int. J. Climatol.* **2005**, *25*, 1965–1978. [CrossRef]
21. Holdridge, L.R.; Grenke, W.C.; Hatheway, W.H.; Liang, T.; Tosi, J.A. *Forest Environments in Tropical Life Zones*; Pergamon Press: New York, NU, USA, 1971; p. 747.
22. Kluge, J.; Kessler, M. Fern endemism and its correlates: Contribution from an elevational transect in Costa Rica. *Divers. Distrib.* **2006**, *12*, 535–545. [CrossRef]
23. Schuur, E.A.G. Productivity and global climate revisited: The sensitivity of tropical forest growth to precipitation. *Ecology* **2003**, *84*, 1165–1170. [CrossRef]
24. Baselga, A. Partitioning the turnover and nestedness components of beta diversity. *Glob. Ecol. Biogeogr.* **2010**, *19*, 134–143. [CrossRef]
25. Koleff, P.; Gaston, K.J.; Lennon, J.J. Measuring beta diversity for presence-absence data. *J. Anim. Ecol.* **2003**, *72*, 367–382. [CrossRef]
26. Harte, J.; Smith, A.B.; Storch, D. Biodiversity scales from plots to biomes with a universal species–area curve. *Ecol. Lett.* **2009**, *12*, 789–797. [CrossRef] [PubMed]
27. Bro, R.; Smilde, A.K. Principal component analysis. *Anal. Methods* **2014**, *6*, 2812–2831. [CrossRef]
28. Chevan, A.; Sutherland, M. Hierarchical Partitioning. *Am. Stat.* **1991**, *45*, 90–96.
29. Mac-Nally, R. Hierarchical partitioning as an interpretative tool in multivariate inference. *Aust. J. Ecol.* **1996**, *21*, 224–228. [CrossRef]
30. Oliver, I.; Nally, R.M.; York, A. Identifying performance indicators of the effects of forest management on ground-active arthropod biodiversity using hierarchical partitioning and partial canonical correspondence analysis. *For. Ecol. Manag.* **2000**, *139*, 21–40. [CrossRef]
31. Mantel, N. The detection of disease clustering and a generalized regression approach. *Can. Res.* **1967**, *27*, 209–220.
32. Jeremy, W.L. Multiple regression on distance matrices: A multivariate spatial analysis tool. *Plant. Ecol.* **2007**, *188*, 117–131.
33. Li, X.W. A floristic study on the seed plants from the region of Yunnan Plateau. *Acta Bot. Yunnanica* **1995**, *17*, 1–14.
34. Fu, C.; Hua, X.; Li, J.; Chang, Z.; Pu, Z.C.; Chen, J.K. Elevational patterns of frog species richness and endemic richness in the Hengduan Mountains, China: Geometric constraints, area and climate effects. *Ecography* **2006**, *29*, 919–927. [CrossRef]
35. Wang, Z.H.; Tang, Z.Y.; Fang, J.Y. Altitudinal patterns of seed plant richness in the Gaoligong Mountains, south-east Tibet, China. *Divers. Distrib.* **2007**, *13*, 845–854. [CrossRef]

36. Wu, Y.; Colwell, R.K.; Han, N.; Zhang, R.; Wang, W.; Quan, Q.; Zhang, C.; Song, G.; Qu, Y.; Lei, F. Understanding historical and current patterns of species richness of babblers along a 5000m subtropical elevational gradient. *Glob. Ecol. Biogeogr.* **2014**, *23*, 1167–1176. [CrossRef]

37. Feng, J.M.; Wang, X.P.; Xu, C.D.; Yang, Y.H.; Fang, J.Y. Altitudinal patterns of plant species diversity and community structure on Yulong Mountains, Yunnan, China. *J. Mt. Sci.-Engl.* **2006**, *24*, 110–116.

38. Sun, Z.H.; Peng, S.J.; Ou, X.K. A quick assessment of tree species richness in Mt. Gaoligong and the environmental interpretation. *Chin. Sci. Bull.* **2007**, *52*, 195–200.

39. Jin, Z.Z. A mossy forest appearing in the sclerophyllous evergreen broad-leaf forests—*Quercetum pannosum nimborum*. Acta Bot. Yunnanica **1981**, *3*, 75–88.

40. Liu, X.L.; He, F.; Fan, H.; Pan, H.L.; Li, M.H.; Liu, S.R. Leaf-form characteristics of plants in *Quercus aquifolioides* community along an elevational gradient on the Balang Mountain in Wolong Nature Reserve, Sichuan, China. *Acta Ecol. Sin.* **2013**, *33*, 7148–7156.

41. Wang, Z.H.; Tang, Z.Y.; Fang, J.Y. The species-energy hypothesis as a mechanism for species richness patterns. *Biodivers. Sci.* **2009**, *17*, 613–624.

42. Hawkins, B.A.; Diniz-Filho, J.A.F. Beyond Rapoport's rule: Evaluating range size patterns of New World birds in a two-dimensional framework. *Glob. Ecol. Biogeogr.* **2006**, *15*, 461–469. [CrossRef]

43. Shen, Z.H.; Fei, S.L.; Feng, J.M.; Liu, Y.N.; Liu, Z.L.; Tang, Z.Y.; Wang, X.P.; Wu, X.P.; Zheng, C.Y.; Zhu, B.; et al. Geographical patterns of community-based tree species richness in Chinese mountain forests: The effects of contemporary climate and regional history. *Ecography* **2012**, *35*, 1134–1146. [CrossRef]

44. Colwell, R.K.; Rahbek, C.; Gotelli, N. The mid-domain effect and species richness patterns: What have we learned so far? *Am. Nat.* **2004**, *163*, E1–E23. [CrossRef] [PubMed]

45. Freestone, A.L.; Inouye, B.D. Dispersal limitation and environmental heterogeneity shape scale-dependent diversity patterns in plant communities. *Ecology* **2006**, *87*, 2425–2432. [CrossRef]

46. Chen, J.; Cao, L.; Chen, Y. Value of world natural heritage 'Three Parallel Rivers'—Scene diversity. *World Nat. Herit.* **2004**, *22*, 22–26.

47. Fjeldså, J.; Álvarez, M.D.; Lazcano, J.M.; León, B. Illicit crops and armed conflict as constraints on biodiversity conservation in the Andes region. *Ambio* **2005**, *34*, 205–211. [CrossRef] [PubMed]

48. Czech, B. Prospects for reconciling the conflict between economic growth and biodiversity conservation with technological progress. *Conserv. Biol.* **2008**, *22*, 1389–1398. [CrossRef] [PubMed]

49. Mutke, J.; Barthlott, W. Patterns of vascular plant diversity at continental to global scales. *Biol. Skr.* **2005**, *55*, 521–531.

forests

MDPI

Article

Lichen Monitoring Delineates Biodiversity on a Great Barrier Reef Coral Cay

Paul C. Rogers [1,2,*], **Roderick W. Rogers** [3], **Anne E. Hedrich** [4] and **Patrick T. Moss** [5]

1   Wildland Resources Department & Ecology Center, Utah State University, Logan, UT 84322, USA
2   Geography Planning and Environmental Management (Visiting Fellow), The University of Queensland, Brisbane 4072, Australia
3   Queensland Herbarium, Brisbane Botanic Gardens, Department of Botany (Emeritus), The University of Queensland, Brisbane 4072, Australia; roderickrogers@westnet.com.au
4   Merrill-Cazier Library, Utah State University, Logan, UT 84322, USA; anne.hedrich@usu.edu
5   Geography Planning and Environmental Management, The University of Queensland, Brisbane 4072, Australia; patrick.moss@uq.edu.au
*   Author to whom correspondence should be addressed; p.rogers@usu.edu; Tel.: +1-435-797-0194.

Academic Editor: Diana F. Tomback
Received: 7 March 2015; Accepted: 28 April 2015; Published: 5 May 2015

Abstract: Coral islands around the world are threatened by changing climates. Rising seas, drought, and increased tropical storms are already impacting island ecosystems. We aim to better understand lichen community ecology of coral island forests. We used an epiphytic lichen community survey to gauge Pisonia (*Pisonia grandis* R.BR.), which dominates forest conditions on Heron Island, Australia. Nine survey plots were sampled for lichen species presence and abundance, all tree diameters and species, GPS location, distance to forest-beach edge, and dominant forest type. Results found only six unique lichens and two lichen associates. A Multi-Response Permutation Procedures (MRPP) test found statistically distinct lichen communities among forest types. The greatest group differences were between interior Pisonia and perimeter forest types. Ordinations were performed to further understand causes for distinctions in lichen communities. Significant explanatory gradients were distance to forest edge, tree density (shading), and Pisonia basal area. Each of these variables was negatively correlated with lichen diversity and abundance, suggesting that interior, successionally advanced, Pisonia forests support fewer lichens. Island edge and presumably younger forests—often those with greater tree diversity and sunlight penetration—supported the highest lichen diversity. Heron Island's Pisonia-dominated forests support low lichen diversity which mirrors overall biodiversity patterns. Lichen biomonitoring may provide a valuable indicator for assessing island ecosystems for conservation purposes regionally.

Keywords: bioindicators; tropical forest; islands; *Pisonia grandis*; *Casuarina equisetifolia*; Australia; ordination; NMS; MRPP; epiphyte

---

1. Introduction

Changing climates are projected to impact low-lying islands as sea levels rise, cyclonic disturbances intensify, and droughts and exotic invasions multiply [1–4]. Plant communities which occupy such locales will likely register the first effects of rapid climate shifts. These forests are also highly prized for their floral and faunal diversity, ecological links to coral reef health, and economic values related to tourism. For example, there are nearly 1000 islands along the Great Barrier Reef (GBR)—a World Heritage Site comprised of coral reefs and cays—that encompass a high priority conservation region paralleling Australia's northeastern coast. Designation of protected areas, of course, is only the first step in a series of actions including gaining ecological insight, designing and implementing plans, monitoring, and adjusting actions toward effective resource conservation. In complex systems, such as coral cay forests, key indicators not only help us to understand relations between ecological components, they provide representational metrics for understanding anthropogenic and natural change.

Pisonia (*Pisonia grandis* R.BR.) forests occur on small islands and coral cays across the Indian and Pacific oceans. Interspecies dynamics of these unique tropical forests are only marginally understood [5, 6]. Throughout the GBR, the greatest concentration of Pisonia forests occurs within the southern Capricorn and Bunker island groups [7]. Coral cays within this region, which are Pisonia-dominant, have been shown to develop a successional pattern where beach grasses and forbs give way to *Casuarina equisitefolia* L. and *Argusia argentea* L.f., which eventually succeeds to recently established Pisonia, then to "old growth" Pisonia farthest from the forest-beach ecotone at an island's interior [5,8]. Pisonia appear to be effective colonizers due to their ability to reproduce by vegetative suckering, rooting of both attached and detached branches, and by seed dispersal and germination. Sticky seeds of Pisonia may become affixed to seabirds and thereby be transported to adjacent islands [5,7,8]. To date, the prime influences on Pisonia forests have been tourist development, drought, invasive insects, and cyclonic disturbance [2–4].

Epiphytic lichens may be used as indicators of broader forest conditions, such as status, health, pollution and other human impacts, and long-term trends [9]. We are unaware of applications of this bioindicator approach in coral cay ecosytems though it is accepted practice elsewhere [9]. Previous studies have linked lichen communities to forest cover change [10,11], wildlife concerns [12], and landscape-level biological diversity [13,14]. Forest systems with a wider range of tree species (*i.e.*, those with diverse bark chemistries and textures) often support broader lichen floras [11,15,16]. In Australia, epiphytic lichens were shown to be particularly sensitive to forest disturbance among a wide suite of indicators [17]. Australia's mainland woodland ecosystems have yielded from 50–60 lichen species [16,18], although little is known about lichen diversity in more monotypic, often geographically isolated, forested island environments.

In the present study we aim to further understand the community ecology of epiphytic lichens on islands dominated by Pisonia forests. Specifically, our objectives are to: (1) document forest conditions and epiphytic lichen flora of Heron Island (Figure 1); (2) determine whether lichen communities differ between dominant forest cover types; and (3) examine causal factors for putative differences in epiphytic lichen communities and how these factors may be affected by shifts in forest cover over time. Greater understanding of coral cay lichens generally, and factors contributing to their diversity and abundance specifically, are expected to inform future conservation efforts where more detailed plant and animal inventories will be cost prohibitive. As coral cay forests change, due to anthropogenic or other factors, lichen inventories, as a proxy for system-wide forest conditions (including biodiversity), may provide an "early alert" avenue for efficient, objective, and credible monitoring.

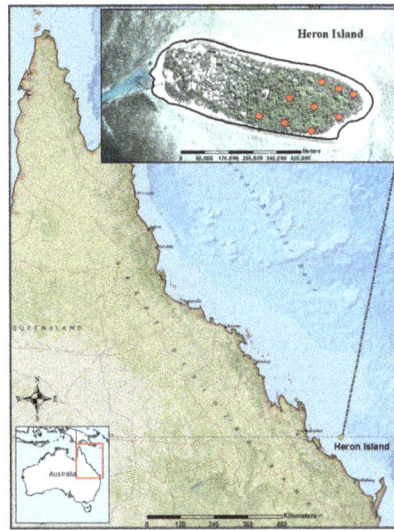

Figure 1. Heron Island study area (inset) in the context of the northeast Australia. Nine lichen monitoring plots (red dots) were selected using a systematic grid overlaid on the Capricornia Cays National Park (eastern) part of the island. The black line surrounding the vegetated portion represents the approximate mean sea level boundary. The *Pisonia grandis* dominated interior section of the forest appears as a denser, slightly lighter green, cover in contrast to the grey-green fringe *Casuarina equisetifolia/Argusia argentea* type. Base map of Heron Island from ©Google 2015 (Imagery date: 2 August 2006).

2. Materials and Methods

2.1. Study Area and Field Methods

Heron Island is a coral cay located on the Great Barrier Reef approximately 80 km northeast of Gladstone, Australia (UTM zone 56 k: 389,285 E, 7,407,039 N). The island is roughly rectangular in shape with an east-west orientation and a high point of just over 9 m. The total area of the island at high tide is about 23 ha, with the vegetated portion covering 20 ha. Annual precipitation averaged 1028 mm and minimum and maximum temperatures were 20.8 and 26.2 °C, respectively, from 1956–2007 (Australian BOM, Heron Island Research Station). About half of Heron Island was excluded from our study due to presence of research, administrative, and tourist facilities. Thus, 9.4 ha of forest terrain on the Capricornia Cays National Park portion (east half) of the island comprised the study landscape. Field activities were conducted during mid-November 2014. Our method set out to establish at least one lichen and forest mensuration plot per h^{-1} of undeveloped forest. We projected a 50 × 50 m grid over the study area and sub-selected sample plots from the alternate intersecting points along roughly east-west grid lines (273° magnetic). The northern most grid line intersected the non-forest (beach) zone, so we reselected a new plot (H2) from the original 50 × 50 m gird (*i.e.*, not an alternating intersection point) same grid line to include a similar edge forest type. Thus, while most sample locations were 100 m from one another, those along the first island transect are just 50 m a part (Figure 1). The final sample design consists of nine plots in three forest types: three each in *Casuarina equisetifolia/Argusia argentea* (CAEQ/ARAR), *Pisonia grandis* mixed (PIGR-mixed), and *Pisonia grandis* dominant (PIGR). This sampling scheme is not proportional to forest type coverage (Pisonia = ~75%–80% of the study area); we favored sampling to maximize substrate and lichen diversity, rather than oversampling what appeared to be a low lichen diversity Pisonia community. Thus, by design, we selected plots along pre-established transect lines, which favored equality among differing forest types, but added some bias against oversampling of Pisonia communities. Since our first objective was to conduct a thorough sample of epiphytic lichens on Heron Island we felt justified in making this decision. In terms of sampling area, each plot represents slightly more than 1 ha of the total forested area within the National Park portion of Heron Island.

For the current study all data was recorded in sub-sampling overlying areas, either 20 m radius (lichen communities) or 7 m radius (tree measures), that were assumed to be representative of a ha^{-1} area centered on grid intersections (plot center). At each plot center we recorded Universal Transverse Mercator (UTM) coordinates using a Geographic Positioning System (GPS) and noted the dominant tree overstorey (forest type). A series of forest measures were taken within a 7 m radius (154 m^2) sample plot. We tallied all trees that reached breast height (1.3 m), noted species and status (live/dead), and recorded their diameter at breast height (dbh) to the nearest cm. Trees were classed by diameter as follows: Mature > 12 cm dbh; Submature = 3–11.9 cm; Immature < 3 cm. Where trees grew in irregular forms at dbh we measured the most consistent narrow portion of the tree bole between basal root collar and multiple trunks [7].

A larger lichen survey area (20 m radius) was centered on the tree plot and followed the basic protocols of the United States Forest Service, Forest Health Monitoring (FHM) program [19,20]. This larger lichen sample area is required to pick up the widest diversity of lichens within a reasonable data collection period. Lichen surveys are conducted under a limited time regime to ensure consistency of sampling between plots. Because our plot areas were much smaller than the FHM program we shortened the survey time to a maximum of 60 min. If no species are found for 10 min following the 30 minute mark the survey is terminated. Lichen field personnel attempt to look at all woody substrates above 0.5 m height, but may include recently fallen tree branches to include lichens which may grow only high in the forest canopy. Putative species are placed in separate packets, given an abundance rating, and positively identified in the laboratory. We included lichen associates—lichen-like bodies of consistent form and relative abundance, but underdeveloped properties (*i.e.*, containing fungal and algal elements)—because we felt they strongly indicate potential for establishment of additional lichen flora. For each lichen packet we recorded tree species substrate, as well as any further identifying characteristics. After completion of the lichen survey each species is assigned a qualitative abundance class for the entire survey area: 1 = 1–3 individuals (distinct thalli); 2 = 4–10 individuals; 3 = more than 10 individuals, but less than presence on 50% of all woody substrates; 4 = presence on more than 50% of woody substrates. A previous study indicated that for sparsely populated epiphytic communities, visual lichen abundance classes were preferable to continuous cover measures because accuracy was comparable while efficiency was greatly increased [21]. Voucher specimens were retained by the lead author, though a number of unknown samples were checked and retained at the Queensland Herbarium, Brisbane Botanic Gardens (second author).

2.2. Derived Variables and Analytical Methods

Following data collection all values were checked for accuracy and completeness, then expanded to reflect the 1 ha sample unit. GPS values were verified for accuracy by projecting them onto a map of Heron Island, then the variable "Distance to Forest Edge" was measured to the nearest meter using GIS (ESRI ArcMap®) digital tools. Trees ha^{-1} were calculated by multiplying the plot tally from the fixed area (154 m^2) by a factor of 66.99. The two larger dbh classes were used to calculate basal area (BA) and the immature class was intended to capture reproduction rates. All ha^{-1} BA values were derived by multiplying the total of all individual tree BAs by same fixed area expansion factor. Live BA and Pisonia BA were calculated separately for each plot to assess their contribution to lichen community diversity. For number of tree species, number of lichens, lichen species abundance, and total lichen abundance we assumed that values obtained on sample plots accurately reflected the larger 1 ha sample unit.

Other than descriptive characterization of the study area forests and lichens (objective 1), our analysis used multivariate statistics to assess potential differences in lichen communities by forest type groups (objective 2) and, if significant differences were found, explored environmental factors contributing to these differences (objective 3). We used PC-ORD® v. 6.0 software [22] for all statistical analyses. Multi-Response Permutation Procedures (MRPP) is a nonparametric test for describing within group agreement of species assemblages. We selected MRPP because of the nature of our small total data matrix and the number of small-occurrence lichen species [23]. The Sørensen distance measure was used because it is less inclined to exaggeration of the outliers inherent in our data set. MRPP produces an *A*-value which is the chance-corrected within group agreement (effect size), as well as a *p*-value establishing level of test significance [24]. For exploratory analyses of explanatory factors we used nonmetric multidimensional scaling (NMS) [25] to ordinate a primary matrix of lichen species by sample locations (plots). A secondary matrix of environmental variables by plots was evaluated in relation to the main species ordination. The lowest stress solution was derived from 250 runs with real plot data. "Stress" is a quantitative assessment final NMS solution monotonicity, a measure of how well real data fit the ordination [23,24]. The lowest stress solution was subjected to a Monte Carlo test of an additional 250 randomized iterations to evaluate the probability of the final NMS solution being greater than chance occurrence. Orthogonal rotation of the final ordination was used to maximize correlation between the strongest environmental variables (*i.e.*, Pearson *r* values) and the major ordination axes. The lowest number of dimensions (axes) was selected when adding another dimension would have decreased the final stress by <5 [24]. For all tests we used a 95% confidence level ($p \leq 0.05$) to determine significance.

3. Results

3.1. Forest Conditions and Lichen Species of Heron Island

Nine forest-lichen sample plots on Heron Island yielded an array of community conditions in a relatively small area. Table 1 presents basic location and forest statistics by three principal forest types. UTM locations simply represent east-west and north-south physical locations of sample plots. Plot distance to forest edge ranged from 8–108 m; those forests farthest from the beach/forest ecotone tended to be nearly pure, often dense, and likely older Pisonia stands. The average number of trees ha^{-1} was 1228, with the greatest number of trees being found on Pisonia and Pisonia mixed plots. Total BA measures were again mostly elevated in Pisonia stands as compared to plots located in forests of non-Pisonia species. Overall, there are very few standing dead trees on Heron Island as shown in the small difference between total and live BA (Table 1). As expected, we find a higher volume of Pisonia BA in the Pisonia forest type *versus* mixed and other types. Two locations (H1, H3) tallied no Pisonia, resulting in overall low BA. The plot the farthest from the forest edge (H15) did not match dominant study patterns: a high number of small trees, largely Pisonia, resulted in a relatively low BA. Pisonia stands contained fewer trees species and lower lichen diversity and abundance (Table 1). We tallied only 12 immature tree stems in the entire study; eight Pisonia and the remaining four divided among *Ficus opposita* Miq. (2), *Cordia subcordata* Lam. (1), and *Celtis paniculata* Endl. (1).

Table 1. Forest statistics by sample location (plot) for Heron Island, Australia. Forest Types are comprised of the principal overstorey tree species: CAEQ/ARAR = *Casuarina equisetifolia/Argusia argentea*; PIGR mixed = 50%–95% *Pisonia grandis* with lesser coverage of CAEQ, ARAR, *Pandanus heronensis* (PAHE), and *Pipturus argenteus* (PIAR); PIGR ≥ 95% PIGR. BA = basal area.

Plot ID	Forest Type	UTM Easting	UTM Northing	Distance to Forest Edge (m)	Trees ha^{-1}	Total BA (m^2/ha^{-1})	Live BA (m^2/ha^{-1})	Pisonia BA (m^2/ha^{-1})	Tree Species Count	Lichen Species Count	Total Lichen Abundance
H1	CAEQ/ARAR	389,458	7,407,117	18.80	780	9.54	9.54	0.00	4	4	14
H2	CAEQ/ARAR	389,405	7,407,147	8.00	325	41.17	40.66	29.39	4	4	12
H3	CAEQ/ARAR	389,368	7,407,153	8.69	390	32.07	31.06	0.00	3	4	14
H11	PIGR mixed	389,467	7,406,983	14.70	975	46.72	46.72	32.52	3	3	10
H13	PIGR	389,387	7,407,023	83.83	1755	144.87	144.74	144.87	1	1	4
H15	PIGR mixed	389,285	7,407,039	103.10	1950	28.92	27.49	15.90	4	2	6
H21	PIGR mixed	389,342	7,406,946	14.36	1820	116.84	116.48	109.29	3	5	12
H23	PIGR	389,261	7,406,963	25.43	1885	97.19	97.19	96.80	2	3	7
H25	PIGR	389,133	7,407,033	31.75	1170	75.91	74.44	74.26	3	1	4

Our survey recorded six identifiable lichens and two lichen "associates" (see Materials and Methods). We included lichen associates—lichen-like bodies of consistent form and relative abundance, but underdeveloped properties—because we felt they strongly indicate potential for establishment of additional lichen flora (both contained fungal and algal elements). Lichen species, as well as their presence, abundance, and prominent substrates are shown in Table 2. The few species tallied parallels a limited diversity in woody substrates on the island. The species list here comprises mostly common northeastern Australian species, with the exception of one incidence of *Strangospora ochrophora* (Nyl.) R.A. Anderson (new to tropical Australia). Lichen forms are exclusively foliose and crustose in the study area, a feature common in environments of limited resources [26]. No individual lichen species was found at every sample location and two species were found only on a single plot. The most common lichen, though not most abundant, was *Hyperphyscia adglutinata* (Flörke) H. Mayrhofer and Poelt. Landscape abundance scores reflect low lichen community presence overall (maximum possible per species = 36). A total of eight tree species were tallied in our survey, though only five of these supported lichens (Table 2). No lichens were recorded on *Ficus opposita*, *Cordia subcordata*, or *Celtis paniculata*, primarily interior forest tree species.

Table 2. Epiphytic lichens and associates recorded by form, frequency, landscape abundance, and substrate tree species. Associates are lichen-like thalli of consistent form and relative abundance on trees in the study area that were unidentifiable due to poor development. Tree species are listed by code: *Argusia argentea* (ARAR), *Casuarina equisetifolia* (CAEQ), *Pandanus heronensis* (PAHE), *Pipturus argenteus* (PIAR), *Pisonia grandis* (PIGR). Landscape abundance is the sum of abundance scores, by species, across all sample plots.

Species	Species Code	Form	Frequency of Presence (% Plots)	Landscape Abundance	Substrate Tree Species
LICHENS					
Dirinaria picta	DIPI	foliose	33	11	CAEQ
Pyxine cocoes	PYCO	foliose	56	19	ARAR, CAEQ, PAHE, PIGR
Hyperphyscia adglutinata	HYAD	foliose	66	17	ARAR, CAEQ, PAHE, PIAR, PIGR
Coenogonium queenslandicum	COQU	crustose	22	3	ARAG, PIGR
Lecanora arthothelinella	LEAR	crustose	11	2	ARAR
Strangospora ochrophora	STOC	crustose	11	2	PIGR
ASSOCIATES					
Cyanobacterium	CYANO	crustose	56	13	ARAR, CAEQ, PIGR
Sterile thalli	THALLI	crustose	44	16	PIGR

3.2. Lichen Community Differences among Forest Types

MRPP results found significant homogeneity within groups for the overall data set ($A = 0.411$, $p = 0.015$), as well as among individual group pairs (Table 3). CAEQ/ARAR *vs.* PIGR displayed the most within group agreement for lichen communities ($A = 0.490$, $p = 0.024$), while the two mixed forest types showed somewhat less similarity though results were highly significant ($A = 0.190$, $p = 0.025$). The lower negative T statistic (-1.364) and insignificant result ($p = 0.09$) for PIGR mixed *vs.* PIGR indicates the weakest between group distinction in lichen tally. Overall, *A* values (*i.e.*, effect size) for this study are somewhat high suggesting a strong group difference, though some caution is warranted given the small sample size [24].

Table 3. Multi-Response Permutation Procedures (MRPP) test results for differences in lichen communities between forest types. "T" is the MRPP test statistic which calculates the difference between observed and expected delta. "A" is the chance-corrected within-group agreement [24] (pp. 188–193).

Forest Type Pairs	T	A	*p*
CAEQ/ARAR *vs.* PIGR mixed	−2.457	0.190	0.025
CAEQ/ARAR *vs.* PIGR	−2.604	0.490	0.024
PIGR mixed *vs.* PIGR	−1.364	0.235	0.097
All types (grand test)	−2.704	0.411	0.015

3.3. Environmental Factors Affecting Coral Cay Lichen Communities

Ordination resulted in a two-dimensional solution on a matrix of eight species by nine sample locations, with a secondary matrix of nine environmental variables. The final NMS solution produced a stress value of 0.001 with an instability of 0.00. A Monte Carlo test of 250 random data runs *versus* the real data set verified a significant NMS outcome ($p = 0.008$). Figure 2 displays a joint plot of the ordination where an overlay of the categorical variable forest type is plotted in lichen "species space" further supporting results of the MRPP test for within group agreement and between group separation. The two-axis solution described about 97% of ordination variance (axis 1: $r^2 = 0.648$; axis 2: $r^2 = 0.339$; orthogonality = 28.5). Length and direction of vectors corresponds to environmental (explanatory) variable strength and relationship to the two-dimensional lichen species space. Only environmental variables making the strongest contributions to species distributions are shown (*i.e.*, $r \geq 0.5$ or < -0.5) in the joint plot (Figure 2). Table 4 presents NMS results by axes for all environmental variables and lichen species. Strong positive and negative responses to axis 1 (Figure 2, Table 4) suggest factors working in opposition to each other in terms of their influence on lichen presence and abundance on Heron Island. As the stronger of the two dimensions represented here, axis 1 appears to describe a gradient of available light, tree density, and tree diversity where higher lichen species richness and abundance align positively with UTM easting and more diverse forest types, and negatively with distance to forest edge and number of trees ha^{-1}. Lack of strong responses along axis 2 indicate poorly defined explanatory factors or the absence of critical elements in our survey (Figure 2). Foliose lichen species (*Dirinaria picta* (Sw.) Clem. and Shear, *Pyxine cocoes* (Sw.) Nyl., *Hyperphyscia adglutinata* (Flörke) H. Mayrhofer and Poelt) strongly favor the apparent light and tree species diversity gradient suggested by axis 1 (Figure 3), while crustose lichens (*Coenogonium queenslandicum* (Kalb and Vězda) Lücking, *Lecanora arthothelinella* Lumbsch (Figure A1), *Strangospora ochrophora* appear to favor the undefined factor(s) determining nearly 40% of the variance within axis 2 of this ordination (Table 4).

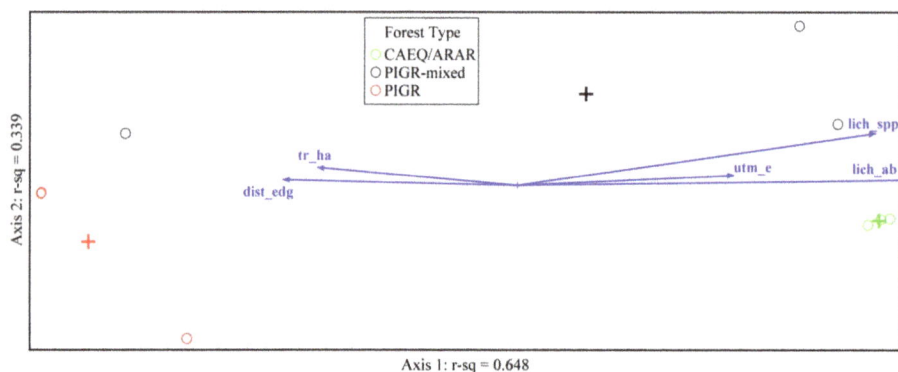

Figure 2. A jointplot depicts the results of nonmetric multidimensional scaling (NMS) ordination of eight lichen species by nine sample plots. Highly correlated environmental variables ($r > 0.5$ or < -0.5; Table 4) are overlaid on the ordination to show relationships to primary axes. Vectors indicate direction (arrow) and strength (length) of these factors in the ordination space defined by plot values of all measured variables. Variables shown are: distance from the sample plot center to forest/beach edge (dist_edg), total trees per ha (tr_ha), lichen species count per plot (lich_spp), UTM location easting (utm_e), and abundance of all lichen species per plot (lich_ab). Forest types are described in Table 1 (CAEQ/ARAR = *Casuarina equisetifolia/Argusia argentea*) and show how groups separate in the Heron Island lichen ordination space. Symbols in bold correspond to forest type group centroid values.

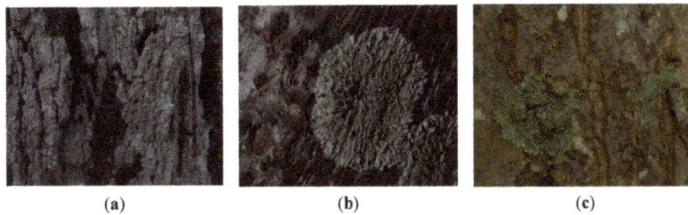

(a)   (b)   (c)

Figure 3. Foliose lichens of Heron Island. (**a**) *Dirinaria picta* on she-oak (*Casuarina equisetifolia*) bark; (**b**) *Pyxine cocoes* on she-oak bark; (**c**) *Hyperphyscia adglutinata* on Pisonia (*Pisonia grandis*) bark. White polygons in 3c are the sterile thalli found commonly throughout the study on Pisonia.

Table 4. Coefficients of determination for correlations between environmental variables, lichen species, and primary ordination axes. Variables in boldface have *r* values >0.5 or <−0.5.

		r Value	
	Code	**Axis 1**	**Axis 2**
	ENVIRONMENTAL VARIABLES		
UTM Easting	utm_e	0.710	0.151
UTM Northing	utm_n	0.401	−0.318
Distance to Forest Edge (beach)	dist_edge	−0.741	0.109
Number of Trees ha^{-1}	tr_ha	−0.685	0.202
Total Basal Area ha^{-1}	ba_total	−0.491	0.111
Live Basal Area ha^{-1}	ba_live	−0.485	0.108
Pisonia grandis Basal Area ha^{-1}	ba_pigr	−0.547	0.072
Number of Tree Species	tr_spp	0.499	0.216
Number of Lichen Species	lich_spp	0.916	0.344
Total Lichen Abundance (Plot)	lich_ab	0.959	0.113
	LICHEN SPECIES		
Dirinaria picta	DIPI	0.667	−0.299
Pyxine cocoes	PYCO	0.974	0.355
Hyperphyscia adglutinata	HYAD	0.978	0.145
Coenogonium queenslandicum	COQU	−0.223	0.544
Lecanora arthothelinella	LEAR	0.261	0.695
Strangospora ochrophora	STOC	−0.308	−0.667
Cyanobacteria (unknown)	CYANO	0.964	0.231
Sterile lichen thalli (unknown)	THALLI	−0.992	−0.319

4. Discussion

4.1. Lichens as Indicators of Forest Diversity

Relatively low plant diversity of Pisonia-dominated islands is mirrored in both the tree and lichen communities of Heron Island. Walker *et al.* [7] recorded 35 vascular plant species for Heron Island, while Batianoff [5] and Batianoff and Hacker [6] documented 40 and 28 plant species for nearby Masthead and Wilson Islands, respectively. We found six identifiable lichen species and two potentially nascent lichen forms within the Heron Island forest community. We acknowledge this low tally of among our target taxon limits the power of statistical analysis, though a species-area curve illustrates that our sampling was adequate for the number of lichen species captured (S2). Also, a very limited lichen flora presents some unique considerations. The greatest diversity of these lichens was found in forests not dominated by Pisonia (Figure A3). Three of the six lichens occurred on only one or two of our nine sample locations and landscape abundance of species was moderate to low (Table 2). Pisonia dominated stands—visually depauperate of understorey vascular plants (Figure 4)—were among the

25

most limited in lichen diversity and total abundance (Table 1). Moreover, the most significant contrast in lichen community make-up was found between CAEQ/ARAR and PIGR forest types, indicating a strong habitat gradient predicated upon the number of tree species present and reflective of greater plant diversity under CAEQ/ARAR cover [5] (Figure A3). We also note an apparent phenotypic distinction between CAEQ/ARAR and PIGR where the former correlated well with foliose lichens and the later favored crustose species (Table 4).

Figure 4. Highly shaded Pisonia (*Pisonia grandis*) forests support few understorey plants or arboreal lichens. Shearwater nests are visible as excavated holes and sandy mounds across the forest floor.

Systematic inventories of coral cay lichen communities are uncommon, but may aid conservationist efforts in rapid assessments of forest status, diversity, and change over time where vascular plant surveys are potentially cost prohibitive. Simple visual surveys, or other techniques such as measures of light penetration, may lack a quantitative basis that relates directly to overall biodiversity such as demonstrated here with a lichen community indicator. In North America [19], Europe [27], and Australia [17] such practices have yielded great insights into forest conditions and are now commonly included in larger suites of national forest monitoring indicators [9,20]. Our challenge will be to assess how well lichen communities track changes in plant diversity as we move from relatively simple island systems to those of greater diversity. For example, Walker *et al.* [7] make a case for higher floral diversity on Pisonia islands of the northern Great Barrier Reef *versus* the southern islands. The bulk of all Pisonia forest coverage in Australia, however, is found in the southernmost island groups [5]. Similarly, we may consider tracking changes in vegetative diversity using successional gradients (based on dominant tree cover) as a surrogate for time [11]. Relationships between biodiversity and Pisonia height, time since disturbance, relative amount of coverage, presence of other tree species, plus salt, light, and wind tolerance have been posited by others [7,28]. Use of this information alongside systematic lichen community surveys can provide indices of plant diversity, as well as establish baseline data for addressing status and trend issues within the broader context of Great Barrier Reef conservation.

4.2. Key Factors Influencing Pisonia-Dominated Lichen Communities

The present study, in both field and analytical approach, was limited in scope and number of lichen species; therefore, results should be viewed in an exploratory vein. Nonetheless, findings presented here suggest that lichen communities, and by extension greater plant diversity, are dependent on relative coverage of Pisonia forests. Near the forest fringe we recorded greater tree diversity and more lichen species; the interior Pisonia forest displayed the opposite pattern (Figure 4). Ordination of all lichen, tree, and environmental data for this study resulted in a strong gradient to support this finding (Figure 2, Table 4). Axis 1 demonstrates a lichen community affinity for available sunlight where those plots located generally further east on the island and having more tree species (COEQ/ARAR) strongly correlated with higher lichen diversity and abundance. In contrast, PIGR and one PIGR mixed tree plot

(H15) were located far from the forest-beach ecotone, had high tree density, and high Pisonia BA, all of which are associated with low light, shaded, and limited tree species environments. Number of tree species is just below the threshold for inclusion in the ordination jointplot, though it is still strongly positively correlated with lichen rich and high light environments (Table 4). In moist forests, available light, often measured in terms of forest gaps, is positively correlated to lichen diversity [10,15]. Further examination of one sample location (H15) is illustrative of several key points. This interior forest stand was dominated by Pisonia, though the presence of three other tree species (*i.e.*, diverse lichen habitat) was insufficient to overcome the negative effects of a low light environment on lichen community vigor (Table 1, Figure 2). Another possible reason for limited lichens at H15 is that recent localized disturbance to this forest has promoted ingrowth of additional trees which are too young to have allowed lichen establishment. This recent "gap" disturbance theory is somewhat supported by the high tree count, but low total BA found at H15.

Axis 2 of our ordination provides no clear explanatory value toward understanding lichen community gradients on Heron Island even though nearly 40% of our variance resides here (Figure 2). One possible explanation that was not specifically tested here is nitrogen (N) deposition. A rich body of research in Europe and North American provides evidence for strong gradients related to airborne N [15,29–31]. In these works, documentation of nitrogen tolerant (nitrophilous) and intolerant lichen species has aided ecologists in determining previously unseen sources of forest degradation. Heron Island, similar to other Pisonia-dominated cays, is a prime breeding site for large populations of two prominent Great Barrier Reef seabirds: white-capped noddy (*Anous minutus* Boie) and wedge-tailed shearwater (*Puffinus pacificus* Gmelin). While the shearwater, which nests underground, is thought to be instrumental in transporting Pisonia seed [7], approximately 70,000 noddies nest in Heron Island tree canopies and deposit an estimated 103 $g \cdot m^{-2}$ of N annually (Figure A4) [32]. Direct deposition of N from seabird guano on trees, as well as indirect uptake of N via groundwater leaching and root uptake may, while apparently being tolerable to Pisonia itself [33], be intolerant for potential lichen colonizers. In the present study, we note that nascent lichens in the form of cyanobacteria were only recorded in locations near the island's perimeter (high light environments) and with putatively low N deposits (non- or mixed-Pisonia forest types). Neitlich and McCune (10) have noted that N-fixing cyanolichens also thrive in high light communities such as those found near Heron Islands forest edge. Additionally, Schmidt *et al.* [33] found a vascular plant gradient of sorts on Heron Island where greater presence of N in plants associated with Pisonia appeared to be lessened in plants at or near the forest-beach ecotone. Further work in this arena will be required to fully understand light, nitrogen, and species diversity issues in coral cay forest environments.

A final factor, also not measured directly here, is tree bark texture. In general, trees with smoother bark are less conducive to lichen establishment [15]. Previous work on North American quaking aspen (*Populus tremuloides* Michx.) tested whether portions of trees with smooth *versus* rough or scarred bark would attract more species and found stark contrast in favor of scarred trunk sections [11]. We note here that Pisonia has smooth bark which may be deleterious to lichen colonization and may be partly responsible for undefined factors reflected in axis 2 of our ordination.

From a temporal perspective, Pisonia-dominated coral cays should be viewed as dynamic systems in which dependent plant communities, including lichens, will respond to periodic forest disturbances and recovery processes. Previous work on Heron Reef (and elsewhere) advocated for an "intermediate disturbance hypothesis" wherein the highest species diversity was associated with moderate disturbance levels [28]. A key explanatory ingredient for lichen communities in the Rocky Mountains, USA, in addition to N deposition was forest succession [31]. In their work, the temporal transition between dominant forest cover types, also an intermediate level, explained more environmental variance than all other factors. Recent work in a boreal setting also supports greater lichen diversity among mixed tree species at mid succession stages [34].

Climate variability plays a key role in Pisonia forest dynamics, with both cyclone activity and drought resulting in tree mortality. There is clear evidence that cyclones can dramatically impact

Pisonia forests, with Cyclone Dinah in 1967 [35], Cyclone David in 1976 [36], and Cyclone Paul in 1980 [3] all damaging Pisonia via wind shearing. These events resulted in much more open forest environments in the 1970s and 1980s compared to the present. Drought can also stress Pisonia trees, making them more susceptible to mortality through infestation by scale insects (*Pulvinaria urbicola* Cockerell) and attendant ants, which collectively reduce Pisonia cover [1]. There is evidence that climate variability may influence scale insects and ant population dynamics, with trees being more susceptible to infestation through stressed trees mobilizing nutrients in the soil during drought events and indirectly through a reduction in Nitrogen soil inputs due to rising sea temperatures reducing prey availability to resident seabirds [2]. Greenslade [2] also noted that scale insect and ant populations dropped when wetter and cooler conditions returned. Thus, we speculate, given the present study addressing lichen communities on Heron Island, that broader regional impacts to Pisonia forests impacted by disturbance and climate change will be reflected in early warning mechanisms such as the lichen bioindicator approach demonstrated here.

5. Conclusions

We conducted a systematic survey of epiphytic lichen communities on a small coral cay on the Great Barrier Reef, Australia. Heron Island, less than 0.2 km², is dominated by *Pisonia grandis* forests which often exhibit low vegetative diversity. The results of this exploratory study strongly suggest that lichen communities are no exception; six identifiable species were confirmed here where mainland forests supported 50–60 lichen species [16,18]. Lichen forms within the forested environment of Heron Island were either foliose or crustose and were most abundant and diverse near the beach-forest ecotone. In contrast, Pisonia-dominated interior forests were nearly depauperate of epiphytic lichens. Conclusions of this study suggest lichens demonstrated distinct preferences for forest communities found near the island's perimeter. The most important explanatory variables for lichen presence, abundance, and distribution on Heron Island were distance to forest edge, number of trees ha^{-1} and Pisonia basal area. As each of these variables increases, they positively relate to the degree of shading and negatively influence lichen occurrence. Mature forests severely limit sunlight penetration, which in turn inhibits most understorey growth. This study highlights a gradient for response to Pisonia shading, providing further evidence for lichens as indicators of broader forest diversity. We speculate that the trends in plant community development shown here will vary depending on frequency of coral cay disturbances. Demonstrated links between lichen communities and forest/successional pathways, such as those shown here, have potential to inform policy and management actions. A key question for future work is whether the lichen biomonitoring techniques applied here are exportable to greater regional studies of island forest development, biodiversity, and change over time.

Acknowledgments: The University of Queensland (UQ), Geography Planning and Environmental Management School provided funding, facilities, and administrative assistance to the lead author under a Visiting Fellowship agreement. UQ Heron Island Research Station staff was instrumental in supporting this research. We wish to thank Queensland Government Department of Environment and Heritage for collections and research permits. Utah State University (USU) Ecology Center and Wildland Resources department also provided logistical support and administration. Chris McGinty, USU College of Natural Resources GIS Lab Assistant Director, provided aid with mapping. We are grateful for the comments of anonymous reviewers.

Author Contributions: Paul C. Rogers, Anne E. Hedrich, and Patrick T. Moss conceived and designed the study; Paul C. Rogers and Anne E. Hedrich implemented the study, collected lichen samples, and compiled/edited field data; Roderick W. Rogers and Paul C. Rogers identified and archived lichen samples; Paul C. Rogers performed all data analysis; Paul C. Rogers, Patrick T. Moss and Anne E. Hedrich wrote the paper.

Conflicts of Interest: The authors declare no conflict of interest.

Forests **2015**, *6*, 1557–1575

Appendix A Supplementary

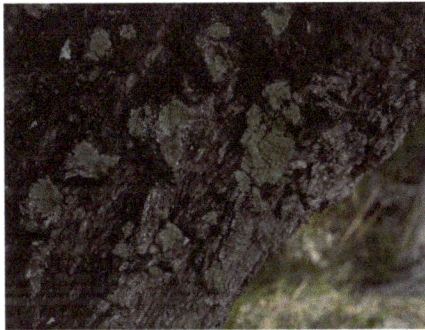

Figure A1. The lichen Lecanora arthothelinella on Argusia argentea bark.

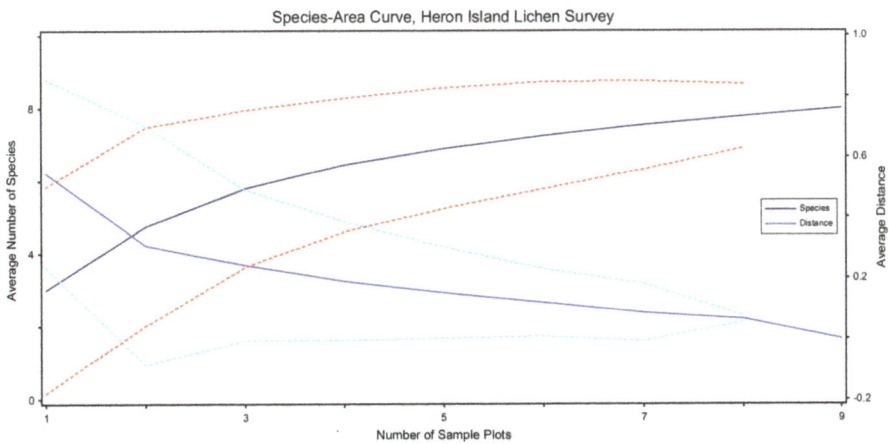

Figure A2. Species-Area Curve for number of plots required to capture complete lichen census at Heron Island, Australia. The asymptotic nature of the species curve (dark blue), as well as Sørensen distance curve (bright blue), describes a maximization of effort (number of plots) required to capture the complete epiphytic lichen flora of the survey area. Dotted lines represent ±1 standard deviation.

(a)

(b)

Figure A3. Two views of island edge forests (CAEQ/ARAR), which allow much greater light penetration, understorey plant cover, and tree diversity. (a) depicts field measures among predominantly *Argusia argentea* cover; (b) shows *Pandanus heronensis*.

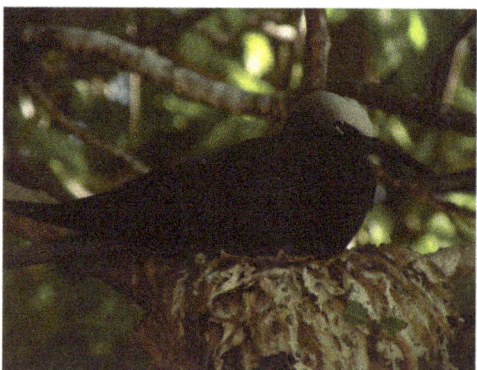

Figure A4. White-capped noddy (*Anous minutus*) nesting in Pisonia (*Pisonia grandis*) tree. The lack of nesting material and great number of noddies present (est. 70,000 birds; [32]) restricts bird nest make-up to only Pisonia leaves.

Forests **2015**, *6*, 1557–1575

References

1. Batianoff, G.N.; Naylor, G.C.; Olds, J.A.; Fechner, N.A.; Neldner, V.J. Climate and Vegetation Changes at Coringa-Herald National Nature Reserve, Coral Sea Islands, Australia 1. *Pac. Sci.* **2010**, *64*, 73–92. [CrossRef]
2. Greenslade, P. Climate variability, biological control and an insect pest outbreak on Australia's Coral Sea islets: Lessons for invertebrate conservation. *J. Insect Conserv.* **2008**, *12*, 333–342. [CrossRef]
3. Ogden, J. On cyclones, *Pisonia grandis* and the mortality of Black Noddy *Anous minutus* on Heron Island. *Emu* **1993**, *93*, 281–283. [CrossRef]
4. Woinarski, J. Biodiversity conservation in tropical forest landscapes of Oceania. *Biol. Conserv.* **2010**, *143*, 2385–2394. [CrossRef]
5. Batianoff, G.N. Floristic, vegetation and shoreline changes on Masthead Island, Great Barrier Reef. *Proc. R. Soc. Qld.* **1999**, *108*, 1–11.
6. Batianoff, G.N.; Hacker, J.L.F. Vascular plant protrait of Wilson Island, Great Barrier Reef, Australia. *Proc. R. Soc. Qld.* **2000**, *109*, 31–38.
7. Walker, T.A.; Chaloupka, M.Y.; King, B.R. Pisonia islands of the Great Barrier Reef. *Atoll Res. Bull.* **1991**, *350*, 1–23. [CrossRef]
8. Cribb, A.B.; Cribb, J.W. *Plant Life of the Great Barrier Reef and Adjacent Shores*; University of Queensland Press: St. Lucia, QLD, Australia, 1985.
9. Nimis, P.L.; Scheidegger, C.; Wolseley, P. (Eds.) *Monitoring with Lichens—Monitoring Lichens*; Kluwer Academic Publishers: Dordrecht, The Netherlands, 2002.
10. Neitlich, P.N.; McCune, B. Hotspots of epiphytic lichen diversity in two young managed forests. *Conserv. Biol.* **1997**, *11*, 172–182. [CrossRef]
11. Rogers, P.C.; Ryel, R.J. Lichen community change in response to succession in aspen forests of the Rocky Mountains, USA. *For. Ecol. Manag.* **2008**, *256*, 1760–1770. [CrossRef]
12. Rosso, A.L.; Rosentreter, R. Lichen diversity and biomass in relation to management practices in forests of northern Idaho. *Evansia* **1999**, *16*, 97–104.
13. Will-Wolf, S.; Esseen, P.A.; Neitlich, P. Monitoring biodiversity and ecosystem function: Forests. In *Monitoring with Lichens—Monitoring Lichens*; Nimis, P.L., Scheidegger, C., Wolseley, P.A., Eds.; Kluwer Academic Publishers: Dordrecht, The Netherlands, 2002; pp. 203–222.
14. Hedenås, H.; Ericson, L. Aspen lichens in agricultural and forest landscapes: The importance of habitat quality. *Ecography* **2004**, *27*, 521–531. [CrossRef]
15. Barkman, J.J. (Ed.) *Phytosociology and Ecology of Cryptogamic Epiphytes*; Van Gorcum & Co.: Assen, The Netherlands, 1958.
16. Morley, S.E.; Gibson, M. Successional changes in epiphytic rainforest lichens: Implications for the management of rainforest communities. *Lichenologist* **2010**, *42*, 311–321. [CrossRef]
17. Abbott, I.; Williams, M.R. Silvicultural impacts in jarrah forest of Western Australia: Synthesis, evaluation, and policy implications of the Forestcheck monitoring project of 2001–2006. *Aust. For.* **2011**, *74*, 350–360. [CrossRef]
18. Stevens, G.N.; Rogers, R.W. The macrolichens flora from the mangroves of Moreton Bay. *Proc. R. Soc. Qld.* **1979**, *90*, 33–49.
19. Will-Wolf, S.; Scheidegger, C.; McCune, B. Methods for monitoring biodiversity and ecosystem function: Monitoring scenarios, sampling strategies and data quality. In *Monitoring with Lichens—Monitoring Lichens*; Nimis, P.L., Scheidegger, C., Wolseley, P.A., Eds.; Kluwer Academic Publishers: Dordrecht, The Netherlands, 2002; pp. 147–162.
20. McCune, B. Lichen communities as indicators of Forest Health. *New Front. Bryol. Lichenol.* **2000**, *103*, 353–356.
21. McCune, B.; Lesica, P. The trade-off between species capture and quantitative accuracy in ecological inventory of lichens and bryophytes in forests in Montana. *Bryologist* **1992**, *95*, 296–304. [CrossRef]
22. McCune, B.; Mefford, M.J. *PC-ORD: Multivariate Analysis of Ecological Data*; MjM Software: Gleneden Beach, OR, USA, 2006.
23. Peck, J.E. *Multivariate Analysis of Community Ecologists: Step-by-Step Using PC-ORD*; MjM Sortware Design: Gleneden Beach, OR, USA, 2010.
24. McCune, B.; Grace, J.B.; Urban, D.L. *Analysis of Ecological Communities*; MjM Software: Gleneden Beach, OR, USA, 2002.

25. Kruskal, J.B. Nonmetric multidimensional scaling: A numerical method. *Psychometrika* **1964**, *29*, 115–129. [CrossRef]

26. Rogers, P.C.; Rosentreter, R.; Ryel, R. Aspen indicator species in lichen communities in the Bear River Range of Idaho and Utah. *Evansia* **2007**, *24*, 34–41. [CrossRef]

27. Wolseley, P.A.; Hill, D.J. Methods for Monitoring Lichens. In *Monitoring with Lichens—Monitoring Lichens*; Nimis, P.L., Scheidegger, C., Wolseley, P.A., Eds.; Kluwer Academic Publishers: Dordrecht, The Netherlands, 2002; pp. 269–272.

28. Connell, J.H. Diversity in tropical rain forests and coral reefs. *Science* **1978**, *199*, 1302–1310. [CrossRef] [PubMed]

29. Van Herk, C.M.; Mathijssen-Spiekman, A.M.; de Zwart, D. Long distance nitrogen air pollution effects on lichens in Europe. *Lichenologist* **2003**, *35*, 347–359.

30. Jovan, S.; McCune, B. Using epiphytic macrolichen communities for biomonitoring ammonia in forests of the greater Sierra Nevada, California. *Water Air Soil Pollut.* **2006**, *170*, 69–93. [CrossRef]

31. Rogers, P.C.; Moore, K.; Ryel, R.J. Aspen succession and nitrogen loading: A case for epiphytic lichens as bioindicators in the Rocky Mountains, USA. *J. Veg. Sci.* **2009**, *20*, 498–510. [CrossRef]

32. Allaway, W.G.; Ashford, A.E. Nutrient input by seabirds to the forest on a coral island of the Great Barrier Reef. *Mar. Ecol. Prog. Ser.* **1984**, *19*, 297–298. [CrossRef]

33. Schmidt, S.; Dennison, W.C.; Moss, G.J.; Stewart, G.R. Nitrogen ecophysiology of Heron Island, a subtropical coral cay of the Great Barrier Reef, Australia. *Funct. Plant Biol.* **2004**, *31*, 517–528. [CrossRef]

34. Bartels, S.F.; Chen, H.Y. Dynamics of epiphytic macrolichen abundance, diversity and composition in boreal forest. *J. Appl. Ecol.* **2015**, *52*, 181–189. [CrossRef]

35. Kikkawa, J. Birds recorded at Heron Island. *Sunbird* **1970**, *1*, 34–48.

36. Flood, P.G.; Jell, J.S. The effect of cyclone 'David' (January 1976) on the sediment distribution patterns on Heron Reef, Great Barrier Reef. In Proceedings of the Third International Coral Reef Symposium, University of Miami, Miami, FL, USA, May 1977; pp. 119–125.

forests

MDPI

Article

Extinction Risk of *Pseudotsuga menziesii* Populations in the Central Region of Mexico: An AHP Analysis

Javier López-Upton [1,†], J. René Valdez-Lazalde [1,†,*], Aracely Ventura-Ríos [1,†], J. Jesús Vargas-Hernández [1] and Vidal Guerra-de-la-Cruz [2]

[1] Graduate School of Forest Sciences, Colegio de Postgraduados, Km. 36.5 Carr. México-Texcoco, Montecillo, Edo. de México 56230, Mexico; uptonj@colpos.mx (J.L.-U.); aracelivr@colpos.mx (A.V.-R.); jjesus.vargashernandez@gmail.com (J.J.V.-H.)

[2] Experimental Site Tlaxcala, INIFAP, Km. 2.5 Carr. Tlaxcala-Santa Ana Chiautempan. Tlaxcala, Tlaxcala 90800, Mexico; guerra.vidal@inifap.gob.mx

* Author to whom correspondence should be addressed; valdez@colpos.mx; Tel.: +52-595-952-0246; Fax: +52-595-952-0256.

† These authors contributed equally to this work.

Academic Editor: Diana F. Tomback

Received: 4 March 2015; Accepted: 27 April 2015; Published: 5 May 2015

Abstract: Within the Analytic Hierarchy Process (AHP) framework, a hierarchical model was created considering anthropogenic, genetic and ecological criteria and sub-criteria that directly affect Douglas-fir (*Pseudotsuga menziesii* (Mirb.)) risk of extinction in central Mexico. The sub-criteria values were standardized, weighted, and ordered by importance in a pairwise comparison matrix; the model was mathematically integrated to quantify the degree of extinction risk for each of the 29 populations present in the study area. The results indicate diverse levels of risk for the populations, ranging from very low to very high. Estanzuela, Presa Jaramillo, Peñas Cargadas and Plan del Baile populations have very low risk, with values less than 0.25. On the other hand, Vicente Guerrero, Morán, Minatitlán, La Garita and Tonalapa populations have very high risk (>0.35) because they are heavily influenced by anthropogenic (close to roads and towns), ecological (presence of exotic species and little or no natural regeneration) and genetic (presence of mature to overmature trees and geographic isolation) factors. *In situ* conservation activities, prioritizing their implementation in populations at most risk is highly recommended; in addition, germplasm collection for use of assisted gene flow and migration approaches, including artificial reforestation, should be considered in these locations.

Keywords: Douglas-fir; risk analysis; conservation of populations; Analytic Hierarchy Process (AHP); multi-criteria evaluation; assisted gene flow

1. Introduction

Pseudotsuga menziesii (Mirb.) Franco barely survived in Mexico after the last ice age; the gradual increase in temperature forced this species to migrate from south to north and towards higher altitude in the mountains [1], which resulted in a fragmented and discontinuous distribution. In Mexico, this conifer is mainly distributed in the northern region, although in the central part of the country the species exists in small isolated stands [2]. At the southernmost limit, two isolated populations are located in the state of Oaxaca [3,4].

Added to this, the improper use and exploitation of natural resources affects forest species to the extent that some of them are endangered or threatened [5,6]. This is the case of *P. menziesii* located in the central region of Mexico, where the species grows in 29 small, fragmented populations suffering high anthropogenic pressure due to land-use changes, overgrazing, forest fires, inappropriate cone collecting, pest attack and illegal tree cutting and timber [7,8]. Also, Mexican populations of *P. menziesii* are exposed to high environmental stress due to their location at the southern end of its natural distribution [7,9]. The above-mentioned situation has caused a reduction in the size and density of populations, as well as low natural recruitment [8]; these circumstances are causing a reduction in genetic diversity and reproductive capacity associated with increased inbreeding [2,10].

P. menziesii is a valuable species for timber and wood production, but in Mexico this conifer is listed as protected by the Mexican government [11], so it is mainly used for Christmas-tree plantations. Therefore, the protection and proper management of each remaining population of the species is important for both natural recovery and establishment of commercial Christmas-tree plantations. It is imperative to implement *in situ* and *ex situ* conservation activities to ensure the preservation of the *P. menziesii* stands in its natural habitat, thereby allowing continuous evolution of the species and conservation of the remaining gene pool.

When designing conservation strategies for the species, we must understand and prioritize the current extinction risk experienced by the remaining populations. Risk of extinction is related to population viability and can be defined as the probability of the continued existence of populations over specified time periods [12], which can be estimated by qualitative and quantitative means. It refers to the likelihood of global or local extinction of target taxa. In this sense these authors suggest that no single measure is sufficient to assess population viability (risk) but different pools of data and type of analysis are necessary. Therefore, several methods to assess population risk might exist based on factors that potentially influence the viability of a species and upon its intrinsic characteristics.

The purpose of this study was to estimate, using a multi-criteria analysis technique, the Analytic Hierarchy Process (AHP), the degree of extinction risk faced by each natural *P. menziesii* population located in Central Mexico (states of Hidalgo, Tlaxcala and Puebla) influenced by anthropogenic, genetic, and environmental factors. Based on AHP, priorities to define a conservation strategy for the species in the region were determined.

2. Materials and Methods

2.1. Analytic Hierarchy Process (AHP)

Ventura-Ríos *et al.* [2] determined the spatial distribution of *P. menziesii* in central Mexico by characterizing 29 small, isolated populations. The present study examines the extinction risk of those populations located in the states of Hidalgo, Tlaxcala and Puebla (Figure 1). The extinction risk analysis was performed using the methodology known as Analytic Hierarchy Process (AHP), a structured approach to complex decision-making defined by Saaty [13] for decision-making in business environments based on the simultaneous analysis of multiple criteria. AHP is one of the most used tools in what is known as multi-criteria decision-making [14]. The tool has been successfully used to define strategies to mitigate the risks associated with decision-making in forestry [15,16] and environmental issues [17]. Theoretical details of the methodology can be found in Saaty [13] and Malczewsky [18]. Saaty and Niemira [14] presented a thorough summary of the technique. We describe the AHP applied to define extinction risk of *P. menziesii* populations located in central Mexico.

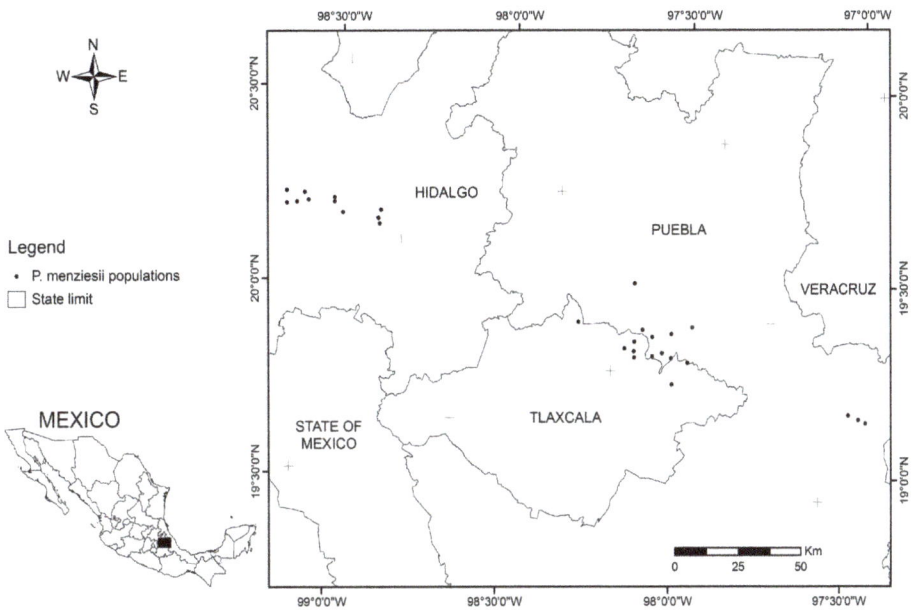

Figure 1. Geographical location of *Pseudotsuga menziesii* (Mirb.) Franco populations in central Mexico.

A model with three hierarchies was developed: definition of goal or objective, identification and definition of criteria for evaluating alternatives, and identification of decision alternatives (Figure 2). Three criteria or decision variables (anthropogenic, genetic and ecological) were considered to estimate the extinction risk of *P. menziesii* populations, each of them composed by several sub-criteria, 14 in total (Figure 2). The three criteria employed are not exhaustive, but these were considered as the most important issues to conduct an assessment of the current risk for these populations.

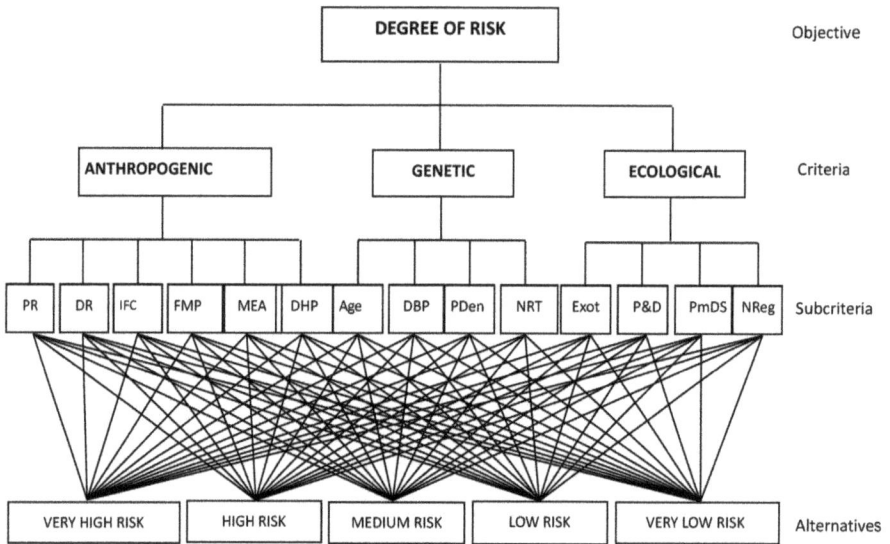

Figure 2. Decision problem hierarchization and sub-criteria considered. PR: Property regime; DR: Distance to roads; IFC: Interest in forest conservation; FMP: Forest management program; MEA: Main economic activity; DHP: Distance to human population; Age: Estimated age of Douglas-fir trees; DBP: Distance between populations of *Pseudotsuga*; PDen: Population density; NRT: Number of trees at reproductive age; Exot: Introduction of exotic species; P & D: Presence of pests and diseases; PmDS: *Pseudotsuga menziesii* is the dominant species; NReg: Natural regeneration.

2.2. Description and Application of Criteria and Sub-Criteria

In order to make a proper comparison using mathematical procedures, sub-criteria were standardized to a common scale or value range from 1 (low contribution) to 9 (high contribution) following the logic for the scale defined by Saaty [13], as shown in Table 1. The value assigned to each sub-criterion, considering the 1-to-9 scale (Table 1), was defined based on scientific literature and considering the results of the survey developed for the owners of *P. menziesii* populations under study (this is indicated following the description of the sub-criteria, according to the case). We used this combination of data sources in order to broaden the perspective of the analysis by including relevant social criteria in the Mexican context. Also, local and site-specific sampling provides more realistic information for the analysis.

Table 1. Standardized values for the sub-criteria used to determine the risk of 29 populations of *Pseudotsuga menziesii* in central Mexico.

DR	Value	MEA	Value	DHP	Value	DBP	Value	NRT	Value
>5	1	Conservation	1	>7.0	1	<2	1	>800	1
3.1–5.0	3	Ecotourism	3	4.1–7.0	3	2–5	3	401–800	3
1.6–3.0	5	Silviculture	5	2.1–4	5	5.1–10	5	201–400	5
0.5–1.5	7	Agriculture—livestock	9	0.5–2.0	7	10.1–30	7	51–200	7
<0.5	9			<0.5	9	>30	9	<50	9
PR		FMP		PDen		Age		NReg	
Communal	1	Yes	1	<10	7	Young	1	Good	1
Ejidal	3	Use rules	3	11–20	5	Mature	4	Low	5
Private	5	No	7	21–30	3	Overmature	4	None	9
				>30	1				
IFC		Exotic		P & D		PmDS			
Yes	1	Yes	7	Yes	7	Yes	1		
No	7	No	1	No	1	Moderate	4		
						No	7		

DR: Distance to roads (km); MEA: Main economic activity; DHP: Distance to human population (km); DBP: Distance between populations of *Pseudotsuga* (km); NRT: Number of trees at reproductive age; PR: Property regimen; FMP: Forest management program and/or forest-use rules; PDen: Population density (No. trees ha^{-1}); Age: Estimated age of Douglas-fir trees; NReg: Natural regeneration; IFC: Interest in forest conservation; Exotic: Introduction of exotic species; P & D: Presence of pest and diseases; PmDS: *Pseudotsuga menziesii* is the dominant species.

2.2.1. Anthropogenic Criteria

Considering that human activities positively or negatively impact the conservation of *P. menziesii* populations, six sub-criteria were included in the criteria (Table 1):

Distance to roads: Distance (km) between the stands (*P. menziesii* populations) and roads; the shorter the distance, the greater the risk of extinction [19].

Property regimen: The type of ownership (communal, private or "ejido") influences the conservation of populations. Based on conducted surveys, it is assumed that forest care is the best when the property regimen is communal, followed by the ejido. Forest in private ownership is the riskiest in the region. An ejido is a land concession given by the Mexican Government to a group of people after the Mexican Revolution (1918); is a collectively owned piece of land in which community individuals possess and farm a specific plot, but forestland is managed on a communal basis. Currently more than 70% of the forestland in Mexico is under this ownership regime.

Interest in forest conservation: If interest is expressed, a value of 1 was assigned; when such interest is lacking a value of 7 was designated. We used such extreme values because the lack of owners' interest in forest conservation makes a great difference in terms of success of a conservation program. This criterion was assessed by the survey taken by owners.

Forest management program (FMP) and/or forest-use rules: Populations that have a FMP approved by the Ministry of Environment and Natural Resources (Secretaría del Medio Ambiente y Recursos Naturales, SEMARNAT) or those under internal rules of management are most likely to be preserved. Data obtained by survey filled out by owners.

Main economic activity: Activities that cause the greatest impact to the forest are agriculture and livestock [20], followed by silviculture, ecotourism and conservation activities. Information based on field observations and survey taken by owners.

Distance to human population: Distance (km) between localities where *Pseudotsuga* grows and human population centers; the greater the distance, the less the risk of possible looting for wood and genetic material [19].

2.2.2. Genetic Criteria

The probability of continued existence is related to the gene pool of populations, because genetic variation helps species to adapt to environmental changes [21,22]. Reduced genetic variability is associated with increased vulnerability in the event of a sudden change in the environment. Four sub-criteria were considered (Table 1).

Estimated age of Douglas-fir trees: The younger the tree, the less the risk. An overmature population is at increased risk because of reduced vigor and decreased reproductive capacity [8]. Trees showing incipient reproductive events and vigorous growth were considered as young, and those that presented evidence of previous reproductive events were considered as mature. These data were obtained by sampling each population [2].

Distance between Pseudotsuga populations (genetic isolation): The longer the distance between populations, the greater the extinction risk. Isolated populations have limited gene flow [23] and genetic isolation may contribute to a reduced genetic variability.

Population density: The lower the density (Number of trees ha^{-1}), the greater the extinction risk. Decreased genetic exchange could generate low genetic variation and limited reproductive capacity, because the presence of few trees of the species leads to diminished pollen density. This criterion was assessed by field sampling.

Number of trees at reproductive age: The lower the number of individuals, the greater the extinction risk. A small population size increases the direct risk of extinction in the presence of a catastrophic event [24]. In addition, a small population has little gene flow and is subjected to random genetic drift, thereby increasing inbreeding, reducing genetic diversity and diminishing good development in the process of species evolution.

2.2.3. Ecological Criteria

Four sub-criteria linked to *P. menziesii* environment and organisms co-inhabiting with the species that could affect its performance and survival were included (Table 1).

Introduction of exotic species: It was considered that populations presenting evidence of *non-native* tree species are most at risk. Exotic species can displace native species by competition for resources [22]. Data evaluated through field surveys.

Presence of pests and diseases: Due to the high impact of cone and seed-feeding insects on Douglas-fir populations in Mexico, evidence that suggests the occurrence of some kind of pest or disease was graded with a higher level of extinction risk; absence of pests and diseases involves lower risk. Data are from field observation.

Pseudotsuga menziesii is the dominant species: The greater the dominance, the less the risk. Dominated species are most likely to disappear due to the strong competition by dominant species. *Pseudotsuga* dominance was determined by field sampling [2].

Natural regeneration: The greater the amount of natural regeneration, the less the risk. Thirty or more seedlings (50 cm in height) per 40 m^2 (average from 3 samples sites) were considered adequate recruitment, whereas less than thirty were considered as scarce. The criterion was assessed by sampling each population [2].

2.3. Construction of Pairwise Comparison Matrices (PCM)

The criteria (Table 2) and sub-criteria (Tables 3–5) were placed in a double entry matrix (called pairwise comparison matrix, PCM) employing the same order for both columns and rows. Subsequently, each pair of criteria and sub-criteria was compared using the 1-to-9 scale defined by Saaty [13]. The cells in the matrices were assessed by comparing the relative importance of the criteria (or sub-criteria) in each row against the criteria (or sub-criteria) listed in each column of the corresponding matrix; due to the principle of reciprocity defined as part of the procedure [13,18], it is only necessary to compare the upper half (above the diagonal) of the PCM. For example, when comparing the importance or contribution of sub-criteria distance to human population (DHP) and distance to roads (DR) for the extinction risk of *Pseudotsuga* populations, it was considered that DHP sub-criteria is moderately more important than DR sub-criteria, so a value of 5 was assigned (Table 3, row 6 and column 2 intersection); in the reciprocal case, the value is 1/5 (Table 3, row 2 and column 6 intersection). All the comparisons listed in Tables 2–5 were similarly performed.

Table 2. Pairwise comparison matrix and relative weights of the criteria used to estimate the degree of risk of the populations of *Pseudotsuga menziesii* in central Mexico [†].

Criteria	Anthropogenic	Genetic	Ecological	*Weight*
Anthropogenic	1	3	3	0.60
Genetic	1/3	1	1	0.20
Ecological	1/3	1	1	0.20

[†] Consistency Ratio (CR) = 0.001; a quantitative measure of consistency associated with the defined pairwise comparisons of criteria.

Calculating the Criteria and Sub-Criteria Weight and Risk Estimation

Estimation of weights for each criterion and sub-criterion considered in the analysis (Tables 2–5) was performed using the algorithm implemented in the WEIGHT module of IDRISI; a software for the analysis and display of spatial data [25]. The weights result from computing a vector of priorities from each matrix; mathematically, the principal eigenvector is computed and normalized [13,14]. The algorithm computes a quantitative measure of consistency, a consistency ratio (CR), associated with the defined pairwise comparisons of criteria and sub-criteria. The CR evaluates the probability that the values in the PCM are not randomly assigned; its value must be less than 0.10, indicating that pairwise comparisons were consistent [13]. In this case, all CR values were lower than 0.10 (Tables 2–5).

Table 3. Pairwise comparison matrix and relative weights of anthropogenic sub-criteria used to estimate the degree of risk of the populations of *Pseudotsuga menziesii* in central Mexico [†].

Anthropogenic Sub-Criteria	PR [‡]	DR	IFC	FMP	MEA	DHP	*Weight*
PR	1	1/3	1/3	1/5	1/9	1/9	0.0301
DR	3	1	1	1/3	1/5	1/5	0.0689
IFC	3	1	1	1/3	1/5	1/5	0.0689
FMP	5	3	3	1	1/3	1/3	0.1532
MEA	9	5	5	3	1	1	0.3394
DHP	9	5	5	3	1	1	0.3394

[†] Consistency Ratio (CR) = 0.017; [‡] PR: Property regime; DR: Distance to roads; IFC: Interest in forest conservation; FMP: Forest management program; MEA: Main economic activity; DHP: Distance to human population.

Table 4. Pairwise comparison matrix and relative weights of genetic sub-criteria used to estimate the degree of risk of the populations of *Pseudotsuga menziesii* in central México [†].

Genetic Sub-Criteria	Age [‡]	DBP	PDen	NRT	*Weight*
Age	1	1/3	1/5	1/7	0.0595
DBP	3	1	1/3	1/5	0.1279
PDen	5	3	1	1/3	0.3639
NRT	7	5	3	1	0.4487

[†] Consistency Ratio (CR) = 0.031; [‡] Age: Estimated age of Douglas-fir trees; DBP: Distance between populations of *Pseudotsuga*; PDen: Population density; NRT: Number of trees at reproductive age.

Table 5. Pairwise comparison matrix and relative weights of ecological sub-criteria to estimate the level of risk of the *Pseudotsuga menziesii* populations' in central Mexico [†].

Ecologic Sub-Criteria	Exot [‡]	P&D	PmDS	NReg	Weight
Exot	1	1/3	1/7	1/7	0.0541
P & D	3	1	1/3	1/3	0.1431
PmDS	7	3	1	1	0.4014
NReg	7	3	1	1	0.4014

[†] Consistency Ratio (CR) = 0.003; [‡] Exot: Introduction of exotic species; P & D: Presence of pests and diseases; PmDS: *Pseudotsuga menziesii* is the dominant species; NReg: Natural regeneration.

To facilitate subsequent arithmetic operations and the interpretation of results by having final values between 0 and 1, standardized values for each sub-criteria (Table 1) were transformed (re-scaled) to values from 0 to 1 (0 = No risk, 1 = Maximum risk). The value of extinction risk for each population was calculated by adding the products of the transformed standardized values of the sub-criteria and their respective weighting values (linear weighted sum). The estimated risk values were ordered from lowest to highest and classified into five categories or degrees of risk: very low (values from 0 to 0.24), low (values of 0.25 to 0.27), medium (values from 0.28 to 0.30), high (values from 0.31 to 0.34) and very high (values greater than 0.35). The risk categories were defined by the authors based on the range of risk values calculated for populations (0.15 to 0.45, Table 6) and the need to clearly prioritize the activities for the conservation of the species. Notice that only one population (Estanzuela) had a risk value below 0.22 (Table 6), in other words, most of the risk values are between 0.22 and 0.45.

Table 6. Degree of risk of extinction for the *Pseudotsuga menziesii* populations in central Mexico.

Population	Value	Risk Level [†]	Population	Value	Risk Level
Estanzuela	0.15	VLow	Barranca Canoita	0.29	Med
Peñas Cargadas	0.22	VLow	Cuyamaloya	0.30	Med
Presa Jaramillo	0.23	VLow	Zapata	0.30	Med
Plan del Baile	0.24	VLow	La Caldera	0.31	High
Cañada El Atajo	0.26	Low	El Llanete	0.32	High
Cruz de Ocote	0.26	Low	Apizaquito	0.33	High
San José Capulines	0.26	Low	La Rosa	0.33	High
Capula	0.27	Low	Las Antenas	0.33	High
Cuatexmola	0.27	Low	Tlalmotolo	0.34	High
El Salto	0.27	Low	Tonalapa	0.36	Vhigh
Villareal	0.27	Low	La Garita	0.38	Vhigh
Axopilco	0.28	Med	Minatitlán	0.42	Vhigh
Buenavista	0.28	Med	Morán	0.44	Vhigh
San Juan	0.28	Med	Vicente Guerrero	0.45	Vhigh
Tlaxco	0.28	Med			

[†] VLow: very low risk; Low: low risk; Med: Medium risk; High: high risk; Vhigh: very high risk.

3. Results

Four of the twenty-nine evaluated populations (13.8%) had very low risk of extinction showing values less than 0.25 (Table 6). These populations, Estanzuela, Peñas Cargadas, Presa Jaramillo and Plan del Baile, are influenced to a lesser degree by anthropogenic factors, because the average distance to human population centers is 3.0 km, the average distance to roads is 3.0 km and the main economic activity for the first three populations is tourism and for the last one is forestry. Moreover, there is interest in preserving populations because the owners of the stands established rules for forest use. *Pseudotsuga menziesii* is the dominant species in these populations, with healthy trees (free of pests and diseases), ages range from young to mature, and presence of natural regeneration from poor to good.

The average distance to the nearest *P. menziesii* population is 4.0 km, the average number of trees at reproductive age is over 500, and the average population density is 23 trees ha^{-1}.

Seven *Pseudotsuga* populations had low risk with values varying from 0.26 to 0.27, including Cañada El Atajo, Cruz de Ocote, San José Capulines, Capula, Cuatexmola, El Salto, and Villareal (Table 6). These populations are influenced by anthropogenic factors because the average distance to human population centers is 2.4 km, the average distance to roads is 2.0 km and the main economic activities are agriculture and silviculture. However, the owners have interest in conserving the populations, since 50% of them have a Forest Management Program and the rest are managed under internal rules for forest use. The *P. menziesii* trees are young, but some populations are infested by pests and diseases. In these locations, Douglas-fir is the dominant species, but its natural regeneration is scarce, the average distance to the nearest *P. menziesii* population is 6.4 km and the average number of trees at reproductive age is 383, while the population density is 18 trees ha^{-1}.

Seven of the *Pseudotsuga* populations were at medium risk with values from 0.28 to 0.30, including Axopilco, Buenavista, San Juan, Tlaxco, Barranca Canoita, Cuyamaloya and Zapata (Table 6). These populations are negatively influenced by anthropogenic and ecological factors. They are at an average distance of 4.2 km away from human population centers and 3.3 km away from roads, but the main economic activity is agriculture. There is interest by owners to conserve the populations, since 50% of these have a Forest Management Program and the others have at least usage rules of the woodlands. Exotic species, pests and diseases are absent; *P. menziesii* trees are mature and dominant, but its natural regeneration is scarce to nonexistent. The average distance to the nearest *P. menziesii* population is 5.4 km, the average number of trees at reproductive age is 263 and the population density is 14 trees ha^{-1}, on average.

Six of the *Pseudotsuga* populations were at high risk, with values between 0.31 and 0.34, including La Caldera, El Llanete, Apizaquito, La Rosa, Las Antenas and Tlalmotolo (Table 6). These populations are strongly and negatively influenced by anthropogenic and genetic factors. They are at an average distance of 3.0 km away from human population centers and 1.8 km away from roads; also the main economic activity is agriculture. The populations are on private land; although some of these have a Forest Management Program and others have rules of use, two of the populations do not have any regulation. Three of these populations showed evidence of pests and diseases. Although *P. menziesii* trees are young and mature, they are not dominant and its natural regeneration is scarce. The average distance to the nearest *P. menziesii* population is 12.8 km, the average number of trees at reproductive age is 299 and the average population density is 11.8 trees ha^{-1}.

The five remaining populations had a very high risk, showing values above 0.35, including Tonalapa, La Garita, Minatitlán, Morán and Vicente Guerrero (Table 6). These populations are strongly and negatively influenced by anthropogenic and genetic factors, with an average distance of 1.2 km away from human population centers and 0.88 km away from roads; in addition, the main economic activity is agriculture. Four of these populations are privately owned and the other is part of an ejido, but none of them has a Forest Management Program. Although these stands varied from young to overmature and they are free from pests and diseases, natural regeneration is lacking and *P. menziesii* is not the dominant species. In this group are the smallest populations like Morán, La Garita and Vicente Guerrero with less than 20 reproductive trees; Minatitlán has 45 individuals, and Tonalapa shows 166 individuals of reproductive age. The average population density is 4.6 trees per ha, and the distance to nearest *P. menziesii* population is greater than 5 km. Furthermore, in these localities exotic pine species have been introduced through reforestation. A special case is the population of Morán, which is at very high risk. This relict of only four reproductive trees was originally described as *Pseudotsuga macrolepis* by Flous [26]. In addition, the owner is considering cutting down the trees because they are growing too close to his home and he is not interested in preserving them.

4. Discussion

The AHP, a mathematical process for measurement and decision-making, was used to evaluate the risk of extinction of fragmented *P. menziesii* populations in central Mexico. It allowed us to propose a hierarchy model that incorporates information from anthropogenic, genetic, and ecological factors or criteria, and offer a reference for future studies on the extinction risk of forest populations. The model was built considering elements that according to scientific literature and the authors experience on forest conservation practices are important to assess the risk of extinction of the species of interest in central Mexico. Neither the model, nor the criteria included in it are exhaustive or can be applied to other situations, but they might be used as a starting point to generate similar models for other tree species present in other latitudes.

AHP facilitated the task of calculating different weight values (importance) for each of the criteria identified as relevant for the risk assessment carried out, a critical issue in the modeling process. Both tasks, hierarchy model development and criteria weighting, are not trivial and must be carefully realized, AHP has widely demonstrated its value to perform such tasks [27,28].

Despite being a mature decision analysis methodology, the AHP is only one of many possible combinations of methods to standardize, weight, and rank available decision alternatives [18]. Other approaches to evaluate extinction risk exist, such as the so called population viability analysis [29]; however, it is also fair to mention that owing to its simplicity, ease of use, and theoretical foundation, the AHP has found wide acceptance among decision-makers [28]. It helps structure the decision problem in a manner that is simple to follow and analyze. Further, it has proved to be a methodology capable of producing results that agree with the decision-maker's expectations; that is the case of the study here reported.

Extinction of native forest tree species is caused by various processes, including reduction of and fragmentation of suitable environment, spread of diseases, genetic erosion and inbreeding [6]. The extinction of species also modifies and alters ecological processes involving it, affecting other species and causing changes in communities and ecosystems [20]. In the case of the *P. menziesii* populations in central Mexico, there is evidence of the effects of fragmentation and isolation by distance (9.4 km average distance between populations) as well as small size, from 4 to 1450 adult trees and 307 trees on average.

It is estimated that coniferous species require a population size of at least 180 reproductive trees to reduce the negative effects caused by low pollen levels and inbreeding, and to maintain seed production at sufficient levels [30]. In several of the stands included in the current study low natural regeneration of *P. menziesii* was observed [8]. In addition, seeds showed low germination and primary dormancy [7,31]. Moreover, a reduced level of genetic diversity with a high proportion of self-fertilization was found [10,32].

Habitat fragmentation in most of these populations is due to anthropogenic factors, but there is evidence that other factors such as environment and genetics play important roles in increasing the risks of extinction. The analysis conducted shows the relative effect of each of these factors over the current risks for each population, which is an important step to outline strategies for species conservation in the region. Risk factors for *Pseudotsuga menziesii* in the study area have their primary origin in the enormous social pressure that persists over populations as a result of historical changes in land use to fulfill human needs in the region, thereby increasing the fragmentation process of habitat.

One way to increase the genetic variability to counteract the effects of inbreeding over reproductive capacity [33,34] is through the promotion of gene flow between neighboring populations growing in similar environments (assisted gene flow as defined by Aitken and Whitlock [35]), which in turn could reduce adaptation problems (outbreeding depression). Populations from similar environments can also exchange genes through assisted migration by creating plantations that use more likely adaptable trees [32]; an action like this one is certainly possible to be carried out in central Mexico, which would increase genetic diversity by transferring new alleles or increasing the frequency of rare alleles in the recipient populations [23,33], and would provide long-term genetic stability to each population and a greater ability for adaptation to climate change or other risk factors. Performing movement of genetic material between related populations would counteract the effects of geographic isolation and fragmentation.

Forests **2015**, *6*, 1598–1612

Conservation of remnant populations of *Pseudotsuga menziesii* in the central region of Mexico should be carried out in the natural environment where the species develops (*in situ*), because this would preserve the natural evolution processes of the species and its interactions with other organisms and ecological processes [22]. Such a conservation strategy is feasible in central Mexico due to the growing interest in managing the populations of the species as a genetic reserve—due to its current legal protection status. In this sense, the Mexican National Forestry Commission (CONAFOR), a federal agency, is currently promoting local and regional programs in order to enhance forest conservation to preserve the country's genetic resources.

The five populations with very high extinction risk and the six populations at high risk should have priority for conservation, due to the real possibility of losing their gene pool in case of extreme events. The populations at medium and low risk would also require protective actions such as grazing prevention, pest control and other actions to reduce current risks. *Ex situ* conservation should be considered as an alternative or complementary resource for populations at high and very high risk.

Other tree species present in central Mexico that are potentially relevant in this context and require studies of extinction risks include: *Pinus chiapensis*, *Fagus sylvatica*, and *Taxus globosa*, all of which have spatial distribution patterns and life history similar to that shown by *Pseudotsuga menziesii*.

5. Conclusions

The AHP-based analysis made it possible to show that 11 of the 29 studied populations were at high risk of extinction, seven were at medium risk and the remaining were at low risk; this suggest different strategies for conservation and protection of *Pseudotsuga menziesii* in the region. Tonalapa, La Garita and Minatitlán populations from the state of Puebla, as well as Morán and Vicente Guerrero from the state of Hidalgo, are at a very high risk of extinction because they are heavily impacted by anthropogenic and genetic factors. These populations are relatively small (less than 200 trees), they are very close to population centers (less than 1.5 km away) and to access roads (less than 1.0 km away) and these populations are located in areas where agriculture is the main economic activity. By contrast, the populations of Estanzuela, Presa Jaramillo, Peñas Cargadas and Plan del Baile are better preserved and they have a very low risk of extinction due to the lower human impacts because populations are at greater distance from human population centers and roads (more than 3.0 km away), in areas where ecotourism is the main economic activity, and some of them are within an ecological reserve.

Acknowledgments: This research was supported by the "Fondo Sectorial CONAFOR-CONACYT (Comisión Nacional Forestal—Consejo Nacional de Ciencia y Tecnología), México", grant 2002-C01-6416. We are grateful for the assistance of two anonymous reviewers that did much to improve the initial manuscript.

Author Contributions: Javier López-Upton and J. René Valdez-Lazalde conceived and designed the study and analyzed data; Aracely Ventura-Ríos generated the data; and all authors contributed in data analysis and discussion. The manuscript was written by J. René Valdez-Lazalde supported by Javier López-Upton and J. Jesús Vargas-Hernández.

Conflicts of Interest: The authors declare no conflict of interest.

References

1. Li, P.; Adams, W.T. Range-wide patterns of allozyme variation in Douglas-fir (*Pseudotsuga menziesii*). *Can. J. For. Res.* **1989**, *19*, 149–161. [CrossRef]

2. Ventura-Ríos, A.; López-Upton, J.; Vargas-Hernández, J.J.; Guerra de la Cruz, V. Caracterización de *Pseudotsuga menziesii* (Mirb.) Franco en el centro de México; Implicaciones para su conservación. *Rev. Fitotec. Mex.* **2010**, *33*, 107–116.

3. Rzedowski, J. *Vegetación de México*; Limusa: México, DF, México, 1978.

4. Del Castillo, R.F.; Pérez de la Rosa, J.A.; Vargas, G.; Rivera, R. Coníferas. In *Biodiversidad de Oaxaca*; García-Mendoza, A.J., Ordóñez, M.J., Briones-Salas, M., Eds.; Instituto de Biología UNAM-Fondo Oaxaqueño para la Conservación de la Naturaleza-World Wildlife Foundation: Mexico, Mexico, 2004; pp. 141–158.

5. CONABIO. *La Diversidad Biológica de México: Estudio de País*; Comisión Nacional para el Conocimiento y Uso de la Biodiversidad: México, DF, México, 1998.

6. Hakoyama, H.; Iwasa, Y.; Nakanishi, J. Comparing risk factors for population extinction. *J. Theor. Biol.* **2000**, *204*, 327–336. [CrossRef] [PubMed]

7. Mápula-Larreta, M.; López-Upton, J.; Vargas-Hernández, J.J.; Hernández-Livera, A. Reproductive indicators in natural populations of Douglas-fir in Mexico. *Biodivers. Conserv.* **2007**, *16*, 727–742. [CrossRef]

8. Velasco-García, M.V.; López-Upton, J.; Angeles-Pérez, G.; Vargas-Hernández, J.J.; Guerra de la Cruz, V. Dispersión de semillas de *Pseudotsuga menziesii* en poblaciones del centro de México. *Agrociencia* **2007**, *41*, 121–131.

9. Hermann, R.K.; Lavender, D.P. *Pseudotsuga menziesii* (Mirb.) Franco. In *Silvics of North America*; Burns, R., Honkala, B.H., Eds.; USDA Forest Service: Washington, DC, USA, 1990; Volume 1, pp. 527–540.

10. Cruz-Nicolás, J.; Vargas-Hernández, J.J.; Ramírez-Vallejo, P.; López-Upton, J. Patrón de cruzamiento en poblaciones naturales de *Pseudotsuga menziesii* (Mirb.) Franco, en México. *Agrociencia* **2008**, *42*, 367–378.

11. SEMARNAT 2010. NOM-059-SEMARNAT-2010, Protección Ambiental-Especies Nativas de México de Flora y Fauna Silvestres-Categorías de Riesgo y Especificaciones Para su Inclusión, Exclusión o Cambio-Lista de Especies en Riesgo. Diario Oficial de la Federación. 30 de Diciembre de 2010. Available online: http://www.profepa.gob.mx/innovaportal/file/435/1/NOM_059_SEMARNAT_2010.pdf (accessed on 10 January 2014).

12. Marcot, B.G.; Murphy, D.D. On population viability analysis and management. In *Biodiversity in Managed Landscapes*; Szaro, R.C., Johnson, D.W., Eds.; Oxford University Press: New York, NY, USA, 1996; pp. 58–76.

13. Saaty, T.L. *The Analytic Hierarchy Process*; McGraw-Hill: New York, NY, USA, 1980.

14. Saaty, T.L.; Niemira, M.P. A framework for making better decisions. *Res. Rev.* **2006**, *13*, 44–48.

15. Bustillos-Herrera, J.A.; Valdez-Lazalde, J.R.; Aldrete, A.; González-Guillén, M.J. Aptitud de terrenos para plantaciones de Eucalipto *Eucaliptus grandis* Hill ex Maiden): Definición mediante el proceso de análisis jerarquizado y SIG. *Agrociencia* **2007**, *41*, 787–796.

16. Olivas-Gallegos, U.E.; Valdez-Lazalde, J.R.; Aldrete, A.; Gonzalez-Guillén, M.J.; Vera-Castillo, G. Áreas con aptitud para establecer plantaciones de maguey cenizo: Definición mediante análisis multicriterio y SIG. *Rev. Fitotec. Mex.* **2007**, *30*, 411–419.

17. Wang, G.; Qin, L.; Li, G.; Chen, L. Landfill site selection using spatial information technologies and AHP: A case study in Beijing, China. *J. Environ. Manag.* **2009**, *90*, 2414–2421. [CrossRef]

18. Malczewski, J. *GIS and Multicriteria Decision Analysis*; John Wiley & Sons: Toronto, ON, Canada, 1999.

19. Vergara, G.; Gayoso, J. Efecto de factores físico-sociales sobre la degradación del bosque nativo. *Bosque* **2004**, *25*, 43–52. [CrossRef]

20. Lambin, E.F. *Modeling Deforestation Processes: A Review*; TREES Series: Research Report No. 1. EUR 15744 EN; European Commission: Luxemburg, Luxemburg, 1994.

21. Soulé, M.E. Conservation: Tactics for a constant crisis. *Science* **1991**, *253*, 744–750. [CrossRef]

22. Primack, R. *A Primer of Conservation Biology*; Sinauer-Sunderland: Sunderland, MA, USA, 1995.

23. Ledig, F.T.; Mápula-Larreta, M.; Bermejo-Velázquez, B.; Reyes-Hernández, J.V.; Flores-López, C.; Capó-Arteaga, M.A. Locations of endangered spruce populations in Mexico and the demography of *Picea chihuahuana*. *Madroño* **2000**, *47*, 71–88.

24. Robledo-Anuncio, J.J.; Alía, R.; Gil, L. Increased selfing and correlated paternity in a small population of a predominantly outcrossing conifer, *Pinus sylvestris*. *Mol. Ecol.* **2004**, *13*, 2567–2577. [CrossRef] [PubMed]

25. Eastman, J.R. *IDRISI Selva Manual*; Clark University: Worcester, MA, USA, 2012.

26. Flous, F. *Diagnoses D'especes et Variétés Nouvelles de Pseudotsuga Americains*; Travaux du laboratorie Forestier de Toulouse: Toulouse, France, 1934; Volume 2, p. 18.

27. Bhushan, N.; Rai, K. *Strategic Decision Making: Applying the Analytic Hierarchy Process*; Springer: London, UK, 2004.

28. Brunelli, M. *Introduction to the Analytic Hierarchy Process*; SpringerBriefs in Operations Research: New York, NY, USA, 2015.

29. Keedwell, R.J. *Use of Population Viability Analysis in Conservation Management in New Zealand*; Science for Conservation 243; New Zealand Department of Conservation: Wellington, New Zealand, 2004.

30. O'Connell, L.M.; Mosseler, A.; Rajora, O.P. Impacts of forest fragmentation on the reproductive success of white spruce (*Picea glauca*). *Can. J. Bot.* **2006**, *84*, 956–965. [CrossRef]

31. Juárez-Agís, A.; López-Upton, J.; Vargas-Hernández, J.J.; Sáenz-Romero, C. Variación geográfica en la germinación y crecimiento inicial de plántulas de *Pseudotsuga menziesii* de México. *Agrociencia* **2006**, *40*, 783–792.

32. Cruz-Nicolás, J.; Vargas-Hernández, J.J.; Ramírez-Vallejo, P.; López-Upton, J. Genetic diversity and differentiation of *Pseudotsuga menziesii* (Mirb.) Franco populations in Mexico. *Rev. Fitotec. Mex.* **2011**, *34*, 233–240.

33. Mosseler, A. Minimum viable population size and the conservation of forest genetics resources. In *Tree Improvement: Applied Research and Technology Transfer*; Puri, S., Ed.; Science Publishers: Enfield, NH, USA, 1998; pp. 191–205.

34. Mosseler, A.; Major, J.E.; Simpson, J.D.; Daigle, B.; Lange, K.; Park, Y.S.; Johnsen, K.H.; Rajora, O.P. Indicators of population viability in red spruce, *Picea rubens*. I. Reproductive traits and fecundity. *Can. J. Bot.* **2000**, *78*, 928–940.

35. Aitken, S.N.; Whitlock, M.C. Assisted gene flow to facilitate local adaptation to climate change. *Annu. Rev. Ecol. Evol. Syst.* **2013**, *44*, 367–388. [CrossRef]

forests

MDPI

Article

How Natural Forest Conversion Affects Insect Biodiversity in the Peruvian Amazon: Can Agroforestry Help?

Jitka Perry [1], Bohdan Lojka [1,*], Lourdes G. Quinones Ruiz [2], Patrick Van Damme [1,3,4], Jakub Houška [1,5] and Eloy Fernandez Cusimamani [1]

[1] Department of Crop Sciences and Agroforestry, Faculty of Tropical AgriSciences, Czech University of Life Sciences Prague, Kamýcká 129, Prague 16500, Czech Republic; jitka.perry@seznam.cz (J.P.); patrick.vandame@ugent.be (P.V.D.); houska@af.czu.cz (J.H.); eloy@ftz.czu.cz (E.F.C.)

[2] CIDRA, Centro de Investigacion y Desarrollo Rural Amazonico, Universidad Nacional de Ucayali, Pucallpa, Peru; lquinonesr@hotmail.com

[3] Laboratory of Tropical and Subtropical Agriculture and Ethnobotany, Ghent University, Coupure links 653, 9000 Ghent, Belgium

[4] World Agroforestry Centre, United Nations Avenue, Gigiri, PO Box 30677, Nairobi 00100, Kenya

[5] Department of Soil Science and Soil Protection, Faculty of Agrobiology, Food and Natural Resources, Czech University of Life Sciences Prague, Kamýcká 129, 16500 Prague, Czech Republic

* Correspondence: lojka@ftz.czu.cz; Tel.: +420-224-382-171; Fax: +420-234-381-829

Academic Editor: Diana F. Tomback
Received: 31 August 2015; Accepted: 23 March 2016; Published: 8 April 2016

Abstract: The Amazonian rainforest is a unique ecosystem that comprises habitat for thousands of animal species. Over the last decades, the ever-increasing human population has caused forest conversion to agricultural land with concomitant high biodiversity losses, mainly near a number of fast-growing cities in the Peruvian Amazon. In this research, we evaluated insect species richness and diversity in five ecosystems: natural forests, multistrata agroforests, cocoa agroforests, annual cropping monoculture and degraded grasslands. We determined the relationship between land use intensity and insect diversity changes. Collected insects were taxonomically determined to morphospecies and data evaluated using standardized biodiversity indices. The highest species richness and abundance were found in natural forests, followed by agroforestry systems. Conversely, monocultures and degraded grasslands were found to be biodiversity-poor ecosystems. Diversity indices were relatively high for all ecosystems assessed with decreasing values along the disturbance gradient. An increase in land use disturbance causes not only insect diversity decreases but also complete changes in species composition. As agroforests, especially those with cocoa, currently cover many hectares of tropical land and show a species composition similar to natural forest sites, we can consider them as biodiversity reservoirs for some of the rainforest insect species.

Keywords: cocoa agroforests; deforestation; diversity losses; habitat conversion; insect diversity; multistrata agroforests; natural forests; species richness

1. Introduction

The Amazon rainforest is considered a biodiversity hotspot [1]. Despite all efforts to conserve and protect this critically important and highly endangered ecosystem, the rainforests still continue to steadily disappear. Over the past 50 years, an estimated 32% of the Amazon rainforest has been converted to man-made systems and a further loss of 10%–15% has been projected by 2050 [2]. Conversion of natural ecosystems is a major cause of biodiversity loss and threatens ecosystem functions and livelihoods of indigenous people [3]. Diversified agricultural systems, such as agroforestry systems, are possibly able to conserve rainforest diversity.

During the last three decades, the Peruvian Amazon ecosystem has been profoundly modified. The extraction of natural resources drives much of the economy. Timber logging, petroleum and natural gas drilling, as well as intense fishing, are amongst the prevailing activities. Forests are the most visibly affected [4]. In Peru, there was an average deforestation rate of about 64,500 ha annually in the period 1999–2005 [5]. Other sources [5–8] further detail deforestation problems in the central part of the Peruvian Amazon, in the area surrounding the city of Pucallpa, where massive deforestation and change of land cover have been observed. Slash-and-burn farming is often blamed as the main cause. Slash-and-burn is a predominant agricultural system in tropical forests in general [9], and the Ucayali region, around Pucallpa city, in particular.

Land use here is quite heterogeneous, with some settlers developing small-scale cattle ranches, while others practice slash-and-burn agriculture (shifting cultivation) with substantial portions of land left fallow [10]. Small-scale farmers often burn down a large part of the forest, simply because they cannot manage the fire. According to Kleinman [11], slash-and-burn farmers transform nutrients stored in forests to soils by slashing, felling, and burning forests. Slash-and-burn is applied mostly in areas where loggers have previously removed the valuable timber trees, whereby the natural forest is then converted into residual or secondary forest. The remaining vegetation is further cut down, burnt and replaced by annual crops, such as rice (*Oryza sativa*), cassava (*Manihot esculenta*), plantain (*Musa* sp.) and/or maize (*Zea mays*), which are frequently planted for both subsistence and local markets. Farmers also have to combat noxious weeds, which often quickly infest degraded lands and pastures. The most problematic weeds are grasses, such as *Imperata brasiliensis* that often develop into green deserts that are periodically burnt down, further spiraling the degradation of the land.

A sustainable and suitable solution that slows down and even halts land degradation is agroforestry, which combines cultivation of trees and crops together. Individual components of this system have different ecological and economic functions. According to De Jong [12], there are various agroforestry systems being practiced in the Peruvian Amazon, near the Ucayali River. Mostly systems combining fruit trees with fast-growing trees earmarked for timber production are found. Guaba (*Inga* sp.) is used by many small-scale farmers as a pilot tree for new agroforestry plots, providing shade and improving the soil condition. It forms a good base for more economically-important species for wood and fruit production. In lowland terrain along the Ucayali River, use of *Inga*, fast-growing timber species bolaina (*Guazuma crinita*) and capirona (*Calycophyllum spruceanum*), in combination with plantains (*Musa* sp.), carambola (*Averrhoa carambola*) and various palm species is very common. Plant species composition depends on locality, economic factors and market opportunities for the farmer. In areas with higher rainfall and more fertile soil in the Ucayali region, the planting of cocoa (*Theobroma cacao*), combined with various fruit and timber trees to provide shade, is gaining popularity.

According to Puri and Panwar [13], agroforestry fields sited close to existing natural forests may benefit from increased biodiversity: for example, animals, birds and insects will be able to invade, and provide important ecosystems services such as pollination, pest control, seed dispersal, *etc.* If there is no such original primary or secondary forest vegetation in close proximity to an agroforestry plot, then the chance of being invaded by a wild species from outside decreases and takes a long time. Degraded lands and pastures in the Pucallpa zone are expanding, and thus the distances between new agroforestry plots and original vegetation are increasing. When the vegetation has already been destroyed or seriously damaged—for example, by unsustainable practices of overgrazing—agroforestry practices may help to restore some biodiversity, either by sheltering relics of the original flora and fauna and allowing them to proliferate, or by creating a new ecosystem, albeit with a different mix of species from that one which originally occupied the site. The most common colonizing species in agroforestry are birds, small vertebrates and invertebrates.

Invertebrates—mainly insects—are ideal indicators for biodiversity, because they are very closely related to the environment, vegetation cover, overall climate, and are sensitive to habitat disturbances. A fast reproductive cycle and a large number of ecological interactions, as well as easy and inexpensive collection, make them the best bio-indicators [14].

As Godfray *et al.* [15] state, understanding insect diversity in the humid tropics is one of the major challenges in modern ecology. Some information sources estimate that tropical forest in the Amazon basin forms a habitat for more than 30% of insect species globally; however, insects also play a significant role in agro-ecosystems found in the Amazon that have not yet been thoroughly studied. Crops grown in conventional monoculture systems often suffer from severe pest problems. This is usually attributed to the nature of the cropping system. Monoculture reduces a complex natural plant system to a single-species community. This can lead to decreased insect diversity and can promote rapid population growth of a single, or very few, insect species [16]. Agroforestry induces a different proportion of various insect species, even herbivorous ones, and this is supposed to lower the occurrence of pest species. Andow [17] conducted studies on 287 species of herbivorous insects. He reports a lower density of herbivore pests in polycultural systems than in monocultural cropping in approximately 52% of the species studied. Only 15% had higher densities. In monocultures in the Peruvian Amazon, there are mainly beetles (most commonly Chrysomelids), Hymenopterans represented by the family *Vespidae*, Homopterans, and others. Species diversity is normally low, but with a great number of individuals for each species. In agroforestry systems, another view on insect diversity, equilibrium and conservation could be offered.

The general objective of our study was to quantify the impact of landuse changes, with respect to diminishing biodiversity, along a disturbance gradient from a tropical forest to a degraded site. Specifically, we set out to evaluate the impact of agroforestry and agricultural practices on insect species composition and diversity in one of the most deforested regions of the Peruvian Amazon. We expected a decrease in insect diversity along the disturbance gradient.

2. Experimental Section

2.1. Study Sites

Several villages in close proximity to Pucallpa city, on the main road to Lima, were chosen as areas, with suitable study sites. The environs of Antonio Raimondi, Pimental and San Alejandro villages (Figure 1) were chosen, and we sampled following landuse systems: multistrata agroforests, cocoa agroforests, cassava monocultures and weedy grasslands on degraded soils (Table 1). As a sample of tropical forest vegetation, we chose the site of Macuya, a natural primary forest of the National University of Ucayali, located close to the Alexander von Humboldt village. The forest was selectively logged more than 20 years ago, displaying the last remnants of human activity. For the purpose of this study, we classified it as a partly logged, yet well-preserved natural forest.

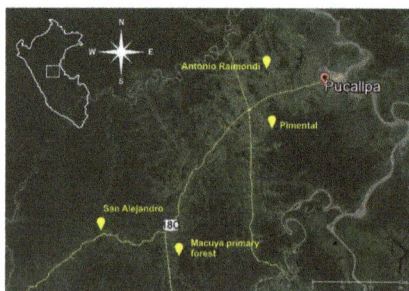

Figure 1. Location of selected study sites near Pucallpa city in the Peruvian Amazon.

Table 1. Study sites characteristics, Peruvian Amazon.

Village	Locality Description	GPS Position	Population	Natural Forest	Multistrata Agroforests	Cocoa Agroforests	Annual Crops	Weedy Vegetation
Macuya	University center for forest research, 12 km from Von Humboldt. Undulated terrain, regular water sources, logged over primary forest, 20+ years old forest on the edge.	S 8°54', W 74°59'	5	Natural forest selectively logged 20+ years, on the edges; trees, palms, shrubs, thick layer of fallen leaves and woody material; Closed canopy; tree density 9 trees/25 m², 95% ground shade, several natural streams and small ponds of standing water.	—	—	—	—
San Alejandro	Undulated with steep slopes, lots of water sources. Local climate allows effective cocoa plantation.	S 8°49', W 75°12'	450	—	—	Cocoa plantation with various shade trees: *Inga* sp., *Calycophyllum* sp., *Dipteryx* sp., *Tabebuia* sp.	Cassava mono-cropping; low vegetation density; no shade	Abundant weeds dominated by *Imperata* grass; compact 1.20 m tall grass vegetation
Pimental	Middle steep terrains and plains, dry with occasional water resources, slash-and-burn farmers, common perennial plantations.	S 8°31', W 74°46'	300	20+ year old forest, closed canopy 5 m, dense vegetation, humid, relatively steep. Small pond of standing water	*Inga edulis*, pineapples, 6 years old, tree density up to 7 trees/25 m²; trees 9 m high; layer of fallen leaves; 80% ground shade, small stream and pond of standing water nearby	*Piper nigrum, Guazuma crinita* six years old; trees 12 m high; vegetation density 6 trees/25 m², 50% ground shade, small stream and pond of standing water nearby	Cassava mono-cropping; vegetation density low; no shade.	Abundant weeds dominated by *Imperata* grass; compact 1.20 m tall grass vegetation
Antonio Raymondi	Slash-and-burn farmers, mainly colonizers. Deforested plains invaded by *Imperata* weeds, eroded soils and dry climate.	S 8°22', W 74°42'	200	—	5 m *Inga edulis*, pineapples, 6 years old, tree density up to 5 trees/25 m²; trees 6 m high; layer of fallen leaves; 75% ground shade	—	Cassava cropping; vegetation density low; no shade.	Plentiful Abundant weeds dominated by *Imperata*; compact 1.20 m tall grass Abundant wminated by *Imperata* grass; compact 1.20 m tall grass vegetation

San Alejandro is a small town, 118 km from Pucallpa (and situated along the Carretera Federico Basadre). The rural residents are dedicated to agriculture (annual crops, livestock, forestry and others). In the last decade, there has been an expansion of cocoa cultivation. For data collection, the village of Porvenir, 2 km from San Alejandro, was chosen. Terrain is diverse, dominated by steep hills with a mosaic of secondary forest vegetation and farm fields. The majority of cropping plots could be classified as agroforestry systems. Sampling was undertaken within two agroforestry plots with shaded cocoa trees.

Pimental is a rural village settlement 35 km from Pucallpa, and 6 km off the main road. Land use systems are based on shifting annual crop cultivation, pepper (*Piper nigrum*) plantations and various fruit tree agroforests. Data were collected within plots with multistrata agroforestry, cassava monoculture and degraded sites with weed vegetation.

Antonio Raymondi village is located 19 km from Pucallpa and 7 km from the main road to Lima, inhabited mainly by small-scale slash-and-burn farmers who cut and burned down large forest areas around the village; they continue to do so to establish new plots for cassava and other staple crops. Some of them apply a multistrata agroforestry system. The village surroundings are degraded lands, covered by the weed species (mainly *Imperata brasiliensis*). Data collection was carried out in two agroforestry plots, cassava monoculture and degraded, weed-infested land. Some of those grasslands had been burned up to three times a year.

2.2. Data Collection

Insects were collected from March to August 2009, and February to November 2010, covering both wet and dry seasons. Sampling was conducted on 12 plots (25 × 25 m), always between 6:30 a.m. and 10:30 a.m., using common entomologic methods. We established four sampling plots in each village. Because of the large distance between villages, each plot was sampled six times, with monthly repetition. Samples in the Macuya forest were taken only four times per sampling period because of logistic problems. On each plot, one Malaise trap, and nine pitfall traps in 5 × 5 m^2 grids were installed; samples were collected after 24 h. As a fix solution for trapped insects, we used a solution of water, salt and soap-like detergent. We also used a sweeping net with a diameter of 40 cm in zigzag transects with a total length of 80 m (one transect per plot).

Each sample was first purified (small soil and plant remnants were taken out), separated into small plastic Ependorf's test tubes or larger laboratory fix-boxes marked by the locality label, and fixed in 70% ethanol.

Samples were sorted into basic taxonomic orders, and determined into specific morphospecies (*i.e.*, distinguished by its morphology from other species). Then the number of species and their relative abundance (number of individuals) were calculated. For biodiversity evaluation, orders of *Coleoptera, Hymenoptera, Orthoptera, Homoptera* and *Dictyoptera* (suborder *Blattodea*) were chosen. Other orders, such as *Diptera, etc.*, were excluded: numerous samples were poorly identifiable under our conditions. These are stored in the depository of the Czech University of Life Sciences, Prague, and are being prepared for further investigation. All the samples were classified with the assistance of professional entomologists.

2.3. Data Evaluation

The goal was to evaluate insect community composition in each ecosystem. Firstly, we documented the insect species composition and their relative abundance in each habitat (landuse system). To account for differences in sample areas, the rarefaction method of Gotelli and Colwell [18] allowed us to construct the species accumulation curves for each habitat based on the number of sampled plots. To show how common or rare a species is in relation to other species in a given location or community, we constructed relative abundance model curves for each locality as a function of abundance and number of species per assessed ecosystem. The Jackknife formula was used to estimate species richness [19] and species heterogeneity was estimated using Shannon-Weiner (H base *e* logs) and Simpson (1-D) diversity indices [19–21] at the level of each sample plot and habitat. For similarity description, Sorensen's index [22] was used. The formula comprises shared species between two habitats and allows us to address the question of how similar the two assemblages are.

To assess statistical differences among the above-mentioned indices and variables of the five land use systems, both parametrical One-way analysis of variance (ANOVA using Tukey's SD test with Bonferroni correction) and non-parametrical Kruskal-Wallis (KW-ANOVA with Bonferroni correction) test were performed for the whole dataset using STATISTICA 9.0 software (StatSoft), as the data show moderate violation of assumptions (normality and heteroscedasticity). However, both approaches give the same results: there are statistically significant differences at least in between some of the groups (habitats). As parametric ANOVA is relatively robust against disruption of its usage assumptions, for *post-hoc* test (multiple comparisons) the Bonferroni test was chosen.

3. Results

3.1. Species Abundance and Composition

We collected insects from a total of 488 traps, from which we selected five orders of insects—*Coleoptera, Hymenoptera, Orthoptera, Homoptera* and *Dictyoptera* (suborder *Blattodea*). Altogether, 64 insect families, represented by 1164 morphospecies, were determined to the species level. In total, we analyzed 3119 individuals. Highest abundance was found in the natural forest, with its 1348 individuals in 758 morphospecies, followed by multistrata agroforests, represented by 992 individuals in 540 morphospecies. Results show a numerical similarity in abundance of cocoa agroforests to monocultures and weed vegetation; however, these localities vary in species composition (Figure 2).

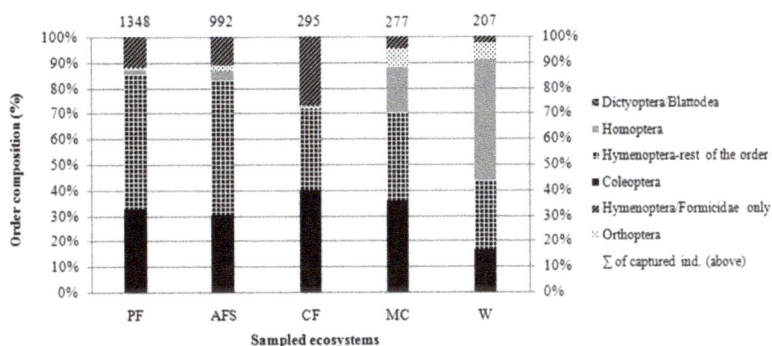

Figure 2. Order-based relative insect composition of five habitats assessed in the Peruvian Amazon (MSF, natural forests; AFS, multistrata agroforests; CF, cocoa agroforests; MC, annual monoculture crops; W, weedy grassland).

Abundance models for each habitat indicate (Supplementary Materials S1) that in natural forest and agroforest habitats, a large number of species are represented by one or two individuals and only a few species are represented by a high number of individuals: this is an indicator of a high species diversity. This pattern occurs in natural non-disturbed ecosystems, or in ecosystems with only low disturbance. On the contrary, a low number of species and their higher abundance (low number of species with only several individuals), reflecting low diversity, can be observed in monoculture habitats and degraded lands infested by weedy grass.

In natural forests, the taxon with the highest number of individuals (relative abundance) was the ants (14% of species samples). The species with the most individuals trapped, was the stingless bee *Trigona pallida* with 41 individuals, and one morphospecies of Chrysomelidae beetles with

40 individuals (family *Chrysomelidae* forms 9% of samples of Macuya forest). In multistrata agroforests, the most abundant morphospecies were all of the *Polistinae* family (47 individuals), followed by a morphospecies of *Cicadelinae* (41) and *Oxybellus* sp., family *Crabronidae* (38 individuals). Overall, the dominant family was *Vespidae*, with a total number of 147 individuals. Cocoa agroforests were dominated by one morphospecies of the *Dolichoderinae* subfamily (*Formicidae*) with 59 individuals (20% of cocoa agroforest samples). The rest of the species in those localities were represented by an average of two individuals per morphospecies (Figure 3). In annual cropping, the most abundant was a morphospecies of digger wasp *Liris* (*Crabronidae*), with 30 individuals. Other species are represented by an average of three individuals. Degraded lands infested by weeds are dominated by a morphospecies of genus *Liris* (15 individuals) and a species of *Agalinae* subfamily (12 individuals). Annual crops and grasslands are characterized by low species richness and abundance (Figure 3). The most abundant species of all the study sites was *Camponatus* sp., an ant present in low numbers in all assessed forest and agroforest ecosystems, yet completely absent in annual cropping and degraded lands. The latter sites are dominated by a morphospecies of digger wasp *Liris*.

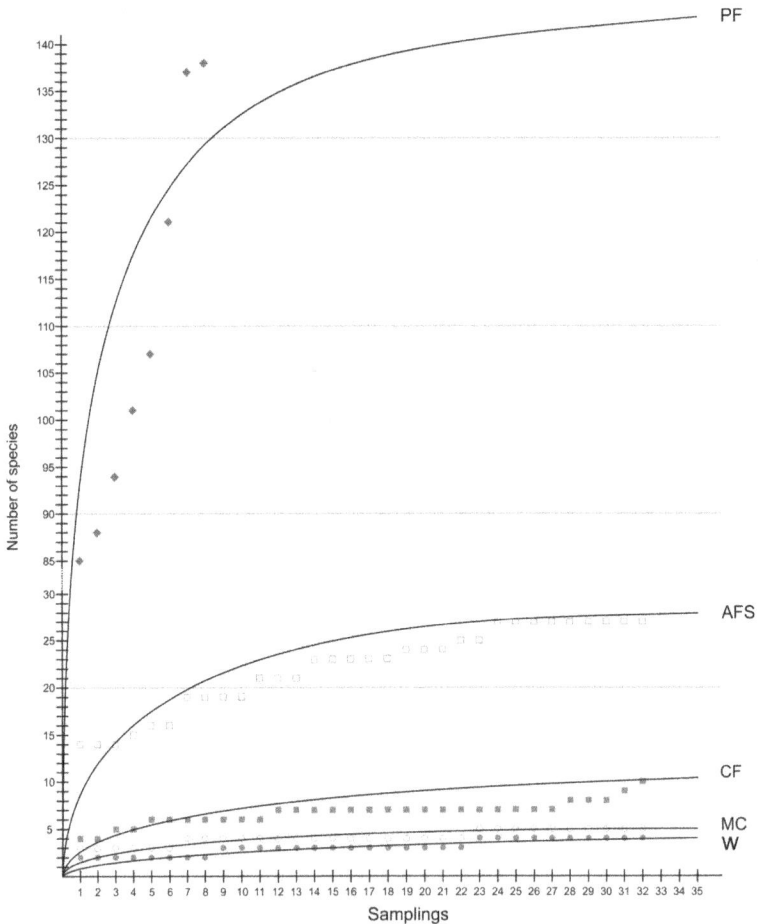

Figure 3. Insect species richness accumulation curves based on number of samples (PF, natural forests; AFS, multistrata agroforests; CF, cocoa agroforests; MC, annual monoculture crops; W, weedy grassland).

3.2. Species Richness and Diversity

To standardize different sample sizes from each of the five habitats, we compared them according to their species accumulation curves (Figure 3). The curve for natural forests based on the number of samples is far from asymptotic, indicating that the area sampled was too small to estimate the total number of species in this land use type. However, with an increased sampling size, insect species richness is likely to be significantly higher in natural forest compared to that of other habitats. It is followed by both agroforestry systems (higher in multistrata agroforests than in cocoa agroforests) and species richness is low but comparable between annual cropping, and weedy grasslands.

To compare relative species richness and diversity among the five habitats (landuse systems), we evaluated the samples in two ways: (i) in total per habitat; and (ii) statistical differences of means per sample and plot (Table 2). In total, all the indices decrease along the human disturbance gradient from undisturbed natural forest to severally disturbed degraded sites covered with weed vegetation (Table 2—total). A comparison of agricultural land use systems indicated, the species richness in cocoa agroforests was more than two times lower than in multistrata agroforests. In annual crop monoculture the species richness was up to six times lower than in multistrata agroforestry and two times lower than in cocoa agroforests.

Statistical evaluation of means per sample and plot indicated significant differences among habitats (Table 2—Means per sample). The results corresponded with the graphic interpretation of data (boxplots and detailed statistical evaluation are presented in Supplementary Materials S2). Both observed species richness and Jackknife estimate of species richness, were highest in natural forest and the values significantly decreased with increasing intensity of land use and/or conversion from multistrata agroforestry towards degraded lands. However, we did not found significant differences among cocoa agroforests, annual cropping and degraded sites with weed vegetation. For species diversity evaluation, the Shannon-Weiner and Simpson's diversity indices were compared. These results also confirm significant decrease in diversity along the disturbance gradient (from natural forests till degraded sites). Shannon-Weiner diversity index reached highest values in natural forests followed by both agroforestry systems (with no significant differences between them) and lowest values (not significantly different) were found in annual cropping and weedy grasslands. Using Simpson's diversity index was highest in natural forests but not significantly different of multistrata agroforests, followed by cocoa agroforests, annual cropping and weedy grasslands (with significant differences among them). The differences between both indices could be explain by the fact that Simpson's index is more weighted towards dominant species compared to Shannon index and those species are more common in disturbed habitats.

Table 2. Insect diversity indices of five habitats in the Peruvian Amazon (sp., species; ind., individuals; sam., sample; Singletons, species with only one individual in samples; Unique sp., species that occur in only one sample, SD, standard deviation). Letters a–d following biodiversity indices values indicate groups significantly differing one from another using Bonferroni post-hoc tests ($\alpha = 0.05$). For detailed statistical results and *post-hoc p*-values, see Supplementary Material S1.

Diversity Indices	Natural Forests	Multistrata Agroforests	Cocoa Agroforests	Annual Crops	Weedy Grasslands	ANOVA p-value
Totals						
Abundance No of ind.	1348	992	297	277	207	
Observed sp. richness No. of sp.	758	540	189	97	64	
Jackknife sp. richness	853	594	203	105	70	
Singletons	157	119	70	24	2	
Unique sp.	126	57	18	11	8	
Shannon-Weiner diversity index	5.21	4.94	4.09	4.07	3.68	
Simpson's diversity index (1-D)	0.991	0.987	0.950	0.973	0.969	
Means per sample $\pm$ SD						
No. of samples	8	31	32	32	32	
Observed sp. richness No. of sp./sam.	109 ± 21.0 a	18.8 ± 3.16 b	6.72 ± 1.22 c	4.13 ± 0.71 c	3.06 ± 0.76 c	<0.001
Jackknife sp. richness	148 ± 28.4 a	24.4 ± 4.10 b	8.38 ± 1.62 c	4.72 ± 1.11 c,d	3.34 ± 0.90 d	<0.001
Sp. density sp./m²	0.121	0.086	0.030	0.016	0.010	
Shannon-Weiner diversity index	0.648 ± 0.155 a	0.181 ± 0.033 b	0.166 ± 0.04 b,c	0.141 ± 0.027 c	0.139 ± 0.029 c	<0.001
Simpson's diversity index (1-D)	0.909 ± 0.031 a	0.883 ± 0.018 a	0.808 ± 0.066 b	0.606 ± 0.081 c	0.554 ± 0.082 d	<0.001

3.3. Similarity among the Habitats

According to Sørensen's index (Table 3), the multistrata agroforestry is highly similar to natural forest (52%), whereby both types share 158 species. Cocoa agroforest shared 34% of species with natural forest and 37% with multistrata agroforests. Monoculture and weedy grasslands shared 30 species and were highly similar (44%). We found only three species shared by all habitats.

Table 3. Similarity between habitats expressed as Sørensen index (in %) and number of shared morphospecies (no. of species) in the Peruvian Amazon. (PF: natural forests; AFS: multistrata agroforests; CF: cocoa agroforests; MC: annual monoculture crops; W: weedy grasslands).

	PF	AFS	CF	MC	W
PF		158	77	51	31
AFS	52%		71	49	26
CF	34%	37%		28	18
MC	24%	28%	27%		30
W	15%	17%	21%	44%	

The similarity relationship is visible on the radar chart graph (spider net—Figure 4), where the majority of ecosystems have their maximum values on the right hand side of the graph (Figure 4), whereas curves of monocultures and weedy grasslands clustered in the left hand side. A "diagonal" difference in species composition can be observed.

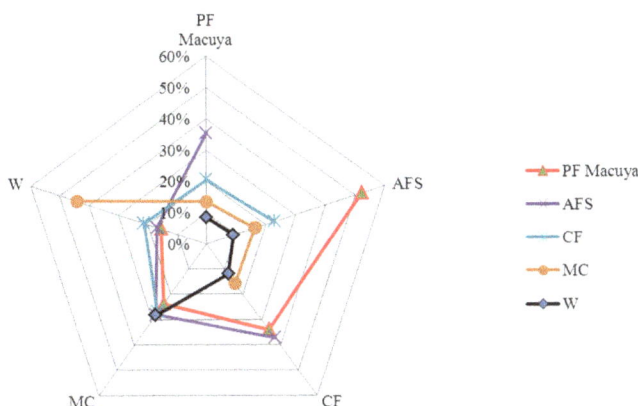

Figure 4. A radar chart (spider net) graph of habitat similarities based on Sørensen index. (PF: natural forests; AFS: multistrata agroforests; CF: cocoa agroforests; MC: annual monoculture cropping; W: weedy grasslands).

4. Discussion

According to our results, the values of species richness and diversity decreased consistently along the disturbance gradient from natural forest towards agroforestry, monoculture crops and degraded lands. This can be explained by the structure of the vegetation cover that is most complex in natural forest. Multistrata agroforests, as the name implies, are characterized by a high diversity of trees, shrubs and annual plants in the undergrowth. They form more complex and suitable environment for various insect species than cocoa agroforests or mono-cropping fields. Cocoa agroforests are mainly cocoa trees shaded by fast-growing timber trees with almost no undergrowth herbaceous plants in them.

Our results are different from those obtained by Bos *et al.* [23], who observed that, compared to other forms of agricultural land use such as annual mono-crops and oil palm plantations, the insect species richness of cocoa agroforests was high and comparable with that of natural forests. Other authors also found cocoa agroforests to be very rich and good biodiversity reservoirs. For example, Rice and Greenberg [24] noted that cocoa plantations can resemble forests in terms of tree cover. Low insect species diversity in this type of vegetation in our study could be caused by specific microclimatic conditions within cocoa plantations, where the undergrowth was nearly missing and herbaceous plant composition was very simple. According to our results, cocoa agroforests are not as high in insect species richness and diversity as multistrata agroforests; however, as cocoa agroforestry systems cover eight million hectares of land worldwide, they have received increased attention for their potential to harbor some of the natural forest biodiversity [25]. According to Bos *et al.* [23], several management practices have an impact on insect richness directly by affecting resources, or indirectly, through creating microclimatic changes and changes in species assemblages. Although there were various factors potentially influencing our results such as proximity of natural water sources in assessed localities and distance from natural forests, we can clearly see that the land use intensity, and ecosystem modification, which is highest in the annual cropping monoculture, highly affects the insect biodiversity and causes visible biodiversity losses and changes in species composition.

Our results on habitat similarity and species composition show a high similarity between agroforestry systems (multistrata and cocoa agroforests) and natural forest. This is supported by Bhagwat [26] who confirmed that heterogeneous agroforestry systems, in which tall trees are maintained and planted for shade (agroforestry systems), form a good refuge for tropical biodiversity. Our results are also consistent with those from other studies [23,27], where the authors claim that, although cocoa agroforests can easily be as rich in insect species as nearby natural forest sites, species assemblages have been found to differ between natural forests and cocoa agroforests, and between differently shaded cocoa agroforests [23].

According to various authors [23,27,28], ants are a rich and ecologically important group of insects that are directly affected by changes in management, for example, by altered resource and nest site availability, but also indirectly, through a complex interplay between microclimate and subsequent changes in species composition. Delabie *et al.* [27] found that a Brazilian cocoa agroforest with a species-poor stand of shade trees harbored a high proportion of forest ants, as compared to other tropical agroecosystems and urban habitats. Similarly, Bos *et al.* [23] found that about half the ant species in Indonesian cocoa agroforests also occurred in nearby forest sites, but species richness decreased with decreasing shade cover [23].

5. Conclusions

Our study presents a detailed analysis of insect biodiversity in several landuse systems in the Ucayali region, Peruvian Amazon. For insect biodiversity, we assessed the natural forest as the most species-rich and diverse, followed by various agroforestry systems. Our results suggest that insect biodiversity may well relate to vegetation structure and complexity. The habitats with least structure, such as annual monocropping and degraded grasslands, showed poor species richness and diversity.

We also found that agroforestry systems possess species composition more similar to natural forest, compared to annual cropping or degraded grasslands. Although species diversity in the latter systems was still relatively high, species composition differed significantly from that in natural forests. These altered and disturbed habitats contained more phytophagous species and a low proportion of predatory species. Higher intensity of land use causes not only a biodiversity decrease, but also a complete change of species composition. As cocoa agroforests nowadays cover large areas of tropical land, there is general recognition that they are biodiversity reservoirs for rainforest species.

Natural forests, as well as all natural vegetation types in the tropics, are worthy of protecting because of their high species richness and ecological potential. The governments and citizens of all tropical countries (not only in Amazon) are responsible for developing agricultural practices that prevent complete ecosystem conversion and loss of the unique biodiversity. We believe that various agroforestry systems, such as multistrata and cocoa agroforests, are able to conserve some of the original biodiversity of original forests, and can form a suitable reservoir for some insect species that occur in primary tropical forests.

Acknowledgments: We would like to thank the professional entomologists Jakub Straka, Štěpán Kubík, Miroslav Barták, Petr Šípek, Petr Janšta, Jakub Straka, and Martin Fikáček as well as the National Museum in Prague. This research was made possible with the financial and scientific support of the Czech Development Cooperation Program (23/Mze/B/07-10), CIGA (Project No. 20145005) and IGA (Project No. 20165016).

Author Contributions: Jitka Perry, was in charge of the whole study (methodology, data collection and evaluation, and writing the manuscript); Bohdan Lojka, the corresponding author, supervised the entire research project, writing of the article, and reviewed and commented on successive drafts of the paper; Lourdes G. Quinones Ruiz supervised data collection and reviewed and commented on successive drafts of the paper; Patrick Van Damme participated in the article writing and reviewed and commented on successive drafts of the paper; Jakub Houška has worked on statistical evaluation of the results and helped with the final editing of the manuscript; and Eloy Fernandez Cusimamani performed data analysis and the interpretation of the results.

Conflicts of Interest: The authors declare no conflict of interest.

References

1. Myers, N.; Mittermeier, R.A.; Mittermeier, C.G.; da Fonesca, G.A.B.; Kent, J. Biodiversity hotspots for conservation priorities. *Nature* **2000**, *403*, 853–858. [CrossRef] [PubMed]

2. Millenium Ecosystem Assessment. *Ecosystems and Human Well-Being: Biodiversity Synthesis*; World Resources Institute: Washington, DC, USA, 2005.

3. Hoekstra, J.M.; Boucher, T.M.; Ricketts, T.H.; Roberts, C. Confronting a biome crisis: Global disparities of habitat loss and protection. *Ecol. Lett.* **2005**, *8*, 23–29. [CrossRef]

4. Clay, J.W.; Clement, C.R. Selected Species and Strategies to Enhance Income Generation from Amazonian Forests. FAO: Misc/93/6 Working Paper: Food and Agriculture Organization of the United Nations. Available online: http://www.fao.org/3/a-v0784e.pdf (accessed on 28 August 2015).

5. Oliviera, P.J.C.; Asner, G.P.; Knapp, D.E.; Almeyda, A.; Galván-Gildemeister, R.; Keene, S.; Raybin, R.F.; Smith, R.C. Land use Allocation Protects the Peruvian Amazon. *Science* **2007**, *317*, 1233–1237. [CrossRef] [PubMed]

6. Hyman, G.; Fujisaka, S. Final Report Desakota, Part II G4. Case Study: Central Peruvian Amazon—A test Case for Desakota Development in the Amazon. International Centre for Tropical Agriculture (CIAT). p. 9. Available online: http://www.dfid.gov.uk/r4d/Output/179310/Default.aspx (accessed on 13 September 2008).

7. Hyman, G.; Puig, J.; Bolanos, S. Multi-source remote sensing and GIS for exploring deforestation patterns and processes in the Central Peruvian Amazon. In Proceedings of the 29th International Symposium on Remote Sensing of Environment, Buenos Aires, Argentina, 8–12 April 2002.

8. Swallow, B.; van Noordwijk, M.; Dewi, S.; Murdiyarso, D.; White, D.; Gockowski, J.; Hyman, G.; Budidarsono, S.; Robiglio, V.; Meadu, V.; *et al.* Opportunities for Avoided Deforestation with Sustainable Benefits. Available online: http://www.asb.cgiar.org/PDFwebdocs/Report-on-Opportunitiesfor-Avoided-Deforestation-Sustainable-Benefits-web-low.pdf (accessed on 6 October 2007).

9. Takasaki, Y. *Economic Model of Shifting Cultivation: A Review*; Tsukuba Economics Working Papers No. 2011-006; University of Tsukuba: Tsukuba, Japan; p. 18.

10. Fujisaka, S.; White, D. Pasture or permanent crops after slash-and-burn cultivation? Land-use choice in three Amazon colonies. *Agrofor. Syst.* **1998**, *42*, 45–59. [CrossRef]

11. Kleinman, P.J.A.; Pimentel, D.; Bryant, R.B. The ecological sustainability of slash-and-burn agriculture. *Agric. Ecosyst. Environ.* **1995**, *52*, 235–249. [CrossRef]

12. De Jong, W.; Freitas, L.; Baluarte, J.; van de Kop, P.; Salazar, A.; Inga, E.; Melendez, W.; Germaná, C. Secondary forest dynamics in the Amazon floodplain in Peru. *For. Ecol. Manag.* **2001**, *150*, 135–146. [CrossRef]

13. Puri, S.; Panwar, P. *Agroforestry: System and Practices*; New India Publishing Agency: New Delhi, India, 2007; p. 657.

14. Brown, K.S., Jr.; Hutchings, R.W. Chapter 7. Disturbance, Fragmentation, and the Dynamics of Diversity in Amazonian Forest Butterflies. In *Tropical Forest Remnants—Ecology, Management and Conservation of Fragmented Communities*; Laurance, W.F., Bierregaard, R.O., Jr., Eds.; The University of Chicago Press: Chicago, UK, 1997; p. 616.

15. Godfray, H.C.J.; Lewis, O.T.; Memmott, J. Studying insect diversity in the tropics. *Philos. Trans. R. Soc. Lond.* **1999**, *354*, 1811–1824. [CrossRef] [PubMed]

16. Stamps, W.; Linit, M. Plant diversity and arthropod communities: Implications for temperate agroforestry. *Agrofor. Syst.* **1997**, *39*, 73–89. [CrossRef]

17. Andow, D.A. Vegetational diversity and arthropod population response. *Annu. Rev. Entomol.* **1991**, *36*, 561–586. [CrossRef]

18. Gotelli, N.J.; Colwell, R.K. Quantifying biodiversity: Procedures and pitfalls in the measurement and comparison of species richness. *Ecol. Lett.* **2001**, *4*, 379–391. [CrossRef]

19. Krebs, C.J. *Ecological Methodology*, 2nd ed.; Addison-Wesley Educational Publishers, Inc.: Boston, MA, USA, 1999.

20. Clarke, K.R.; Warwick, R.M. *Changes in Marine Communities: An Approach to Statistical Analysis and Interpretation*, 3rd ed.; PRIMER-E: Plymouth, UK, 2001; p. 260.

21. Khan, A. Methodology for Assessing Biodiversity. Centre of Advanced Study in Marine Biology, Annamalai University. Available online: http://www.slideshare.net/MMASSY/methodology-forassessmentbiodiversity (accessed on 28 August 2015).

22. Sörensen, T.A. A method of establishing groups of equal amplitude in plant sociology based on similarity of species content, and its aplication to analyses of the vegetation on Danish commons. *K. Dan. Vidensk. Selsk.* **1948**, *5*, 1–34.

23. Bos, M.M.; Steffan-Dewenter, I.; Tscharntke, T. The contribution of cacao agroforests to the conservation of lower canopy ant and beetle diversity in Indonesia. *Biodivers. Conserv.* **2007**, *16*, 2429–2444. [CrossRef]

24. Rice, R.A.; Greenberg, R. Cacao cultivation and the conservation of biological diversity. *Ambio* **2000**, *29*, 167–172. [CrossRef]

25. Urquhart DH. *Cocoa*; Longmans Green and Co: London, UK, 1961.

26. Bhagwat, S.A.; Willis, K.J.; Birks, H.J.B.; Whittaker, R.J. Agroforestry: A refuge for tropical biodiversity? *Trends Ecol. Evol.* **2008**, *23*, 261–267. [CrossRef] [PubMed]

27. Delabie, J.H.C.; Jahyny, B.; Cardoso do Nascimento, I.; Mariano, C.S.F. Contribution of cocoa plantations to the conservation of native ants (*Insecta: Hymenoptera: Formicidae*) with a special emphasis on the Atlantic Forest fauna of southern Bahia, Brazil. *Biodivers. Conserv.* **2007**, *16*, 2359–2384. [CrossRef]

28. Ambrecht, I.; Perfecto, I.; Vandermeer, J. Enigmatic biodiversity correlations: Ant diversity responds to diverse resources. *Science* **2004**, *304*, 284–286. [CrossRef] [PubMed]

forests

Article

Outbreak of *Phoracantha semipunctata* in Response to Severe Drought in a Mediterranean *Eucalyptus* Forest

Stephen Seaton [1,*], George Matusick [1,2,†], Katinka X. Ruthrof [1,†] and Giles E. St. J. Hardy [1,†]

[1] Centre of Excellence for Climate Change Woodland and Forest Health, School of Veterinary and Life Sciences, Murdoch University, Murdoch, Western Australia 6150, Australia; g.matusick@murdoch.edu.au (G.M.); k.ruthrof@murdoch.edu.au (K.X.R.); g.hardy@murdoch.edu.au (G.E.S.J.H.)

[2] The Nature Conservancy, Georgia Chapter, Chattahoochee Fall Line Conservation Office, Fort Benning, GA 31905, USA

* Author to whom correspondence should be addressed; s.seaton@murdoch.edu.au; Tel.: +61-893-606-272; Fax: +61-893-606-303.

† These authors contributed equally to this work.

Academic Editor: Diana F. Tomback

Received: 29 August 2015; Accepted: 27 October 2015; Published: 30 October 2015

Abstract: Extreme climatic events, including droughts and heatwaves, can trigger outbreaks of woodboring beetles by compromising host defenses and creating habitat conducive for beetle development. As the frequency, intensity, and duration of droughts are likely to increase in the future, beetle outbreaks are expected to become more common. The combination of drought and beetle outbreaks has the potential to alter ecosystem structure, composition, and function. Our aim was to investigate a potential outbreak of the native *Eucalyptus* longhorned borer, *Phoracantha semipunctata* (*P. semipunctata*), following one of the most severe droughts on record in the Northern Jarrah Forest of Southwestern Australia. Beetle damage and tissue moisture were examined in trees ranging from healthy to recently killed. Additionally, beetle population levels were examined in adjacent forest areas exhibiting severe and minimal canopy dieback. Severely drought-affected forest was associated with an unprecedented outbreak of *P. semipunctata*, with densities 80 times higher than those observed in surrounding healthier forest. Trees recently killed by drought had significantly lower tissue moisture and higher feeding damage and infestation levels than those trees considered healthy or in the process of dying. These results confirm the outbreak potential of *P. semipunctata* in its native Mediterranean-climate *Eucalyptus* forest under severe water stress, and indicate that continued drying will increase the likelihood of outbreaks.

Keywords: woodborer; damage; die-off; die-back; forest mortality; heat wave; *Cerambycidae*; populations; climate change; Jarrah

---

1. Introduction

Woodboring beetles are among the most widespread insect pests of trees, and can be especially responsive to drought stress in their hosts [1–3]. With droughts expected to increase with climate change in many regions [4], outbreaks of some woodboring beetles are expected to become more frequent and intense [5]. For example, forests in Mediterranean regions are already experiencing significant shifts in precipitation and temperatures, increasing tree water stress [6–8]. Associated with these climate changes, forests are being challenged from abiotic factors (e.g. high temperatures and drought) and by biotic factors such as damaging outbreaks of woodboring beetles in semiarid, temperate, and boreal climate regions [9–12]. While bark beetles (*Coleoptera: Curculionidae*) are well-known forest pests of drought-stressed hosts, other woodboring beetles (e.g., *Cerambycidae* and *Buprestidae*) can also cause widespread damage under drought conditions [13]. For example, in the Ozark mountains (a temperate climate region) of the Southeastern United States, outbreaks of the red oak borer (*Enaphalodes rufulus* Haldeman (*E. rufulus*) (*Coleoptera: Cerambycidae*)), coinciding with drought and loss of tree vigor, led to large-scale mortality of several *Quercus* species in the early 2000s [13].

Phoracantha semipunctata (*P. semipunctata*) (*Coleoptera: Cerambycidae*) has become a major pest of *Eucalyptus* species throughout the world where they have been introduced [14,15]. Worldwide, it is among the most economically damaging *Phoracantha* species [15]. For example, in California, it causes extensive tree losses, attacking a wide range of *Eucalyptus* species in plantations [16,17]. Drought stress is a predisposing factor for *P. semipunctata* attack and tree mortality [15]. Despite the amount of literature on this species outside of Australia, in its native forest ecosystems in drought prone Australia, where the insect has a wide distribution [18], its behavior and response to drought stress in its hosts has received limited attention. In fact, evidence of *P. semipunctata* causing widespread damage to its native ecosystems is rare [19,20].

The Mediterranean climate region of Southwestern Australia has experienced a climatic shift since the 1970s, with a step-wise reduction in rainfall and steady increase in average temperature [5]. Coinciding with this drying trend, a series of canopy dieback events have occurred in multiple ecosystems over the past two decades [21,22], with some associated with outbreaks of endemic woodboring beetles [23]. The largest, in terms of spatial extent, and most damaging event, occurred during the Australian summer 2010/2011 in the largest forest type in the region (the Northern Jarrah Forest) [24]. While dieback and tree mortality were largely suspected to be driven by the combination of drought and heat, woodboring beetle activity was also evident in many dying and recently-killed trees [24]. Collection of adults from infested stems showed that the native *P. semipunctata* was the primary beetle present in the two overstory dominant tree species in the Northern Jarrah Forest, *Eucalyptus marginata* Donn ex Smith (*E. marginata*) and *Corymbia calophylla* L.A.S Johnson (*C. calophylla*) (Matusick, pers. observations). Although the behavior of *P. semipunctata* is well known where it has been introduced outside of Australia [15,16,25], little research has been conducted on its population biology and host damage in native ecosystems. Furthermore, given the close association of *P. semipunctata* with tree water stress, and the magnitude of climate changes projected for the Australian continent, research on this species is particularly important.

Following the cessation of drought in 2011, we conducted a study to determine the host damage and infestation response of *P. semipunctata* in relation to drought stress in the Northern Jarrah Forest of Southwestern Australia. Two surveys were conducted (2011 and 2012) to determine (1) the association between tree crown health and damage by *P. semipunctata* (2011), and (2) the infestation of *P. semipunctata* in severely and minimally drought-affected forest areas and among trees of varying health conditions (2012). Results will determine the strength of the association between *P. semipunctata* and severe host drought stress, and provide a basis to estimate its outbreak potential with continued drying in the region.

2. Experimental Section

2.1. Study Area

The Northern Jarrah Forest ranges from tall open sclerophyll forest in the west to open woodland in the east due to a strong west-east precipitation (ranging from 1300 mm to 800 mm) and evaporation gradient (ranging from 575 mm to 700 mm) [26]. The uplands are dominated by *E. marginata* and *C. calophylla*, which commonly occur as overstory codominants. Common midstory species include *Allocasuarina fraseriana* (Miq.) L.A.S.Johnson, *Banksia grandis* Wild, *Persoonia longifolia* R.Br, and *Persoonia elliptica* R.Br [27]. Soils throughout the study area are lateritic, predominantly ironstone gravels with sandy matrixes of low moisture holding capacity, and include many areas of shallow soils over granite [28].

2.2. Study Site and Plot Selection

Twenty sites were selected for field sampling from a population of 235 severely drought-affected sites determined by an aerial survey in May 2011 (Figure 1). For detailed methods of the aerial survey see Matusick *et al.* [24]. Briefly, the severely-affected sites (>70% complete crown discoloration) were located from the aerial survey and accurately delineated using a differential GPS (Pathfinder Pro XRS receiver, Trimble Navigation Ltd., Sunnyvale, CA, USA) (±0.4 m) on the ground.

In June 2011, shortly following the cessation of drought, a survey was conducted to estimate beetle damage to trees. Plots were selected by choosing 3 points randomly on a 20 m × 20 m grid (using fGIS forestry cruise software (Wisconsin DNR-Division of Forestry, Madison, United States) from within the severely-affected forest in each of the 20 sites selected and represented the center point of a 0.011 ha fixed radius forest plot. Severely-affected areas were associated with shallow soils surrounding rock outcrops and on soil types with lower water holding capacity than minimally-affected areas [29]. Additionally, 3 plots were located outside the severely-affected area in forest minimally-affected (<20% crown discoloration). These plots were located 20 m outside the perimeter of the severely-affected area. A differential GPS was used to locate plots within each area.

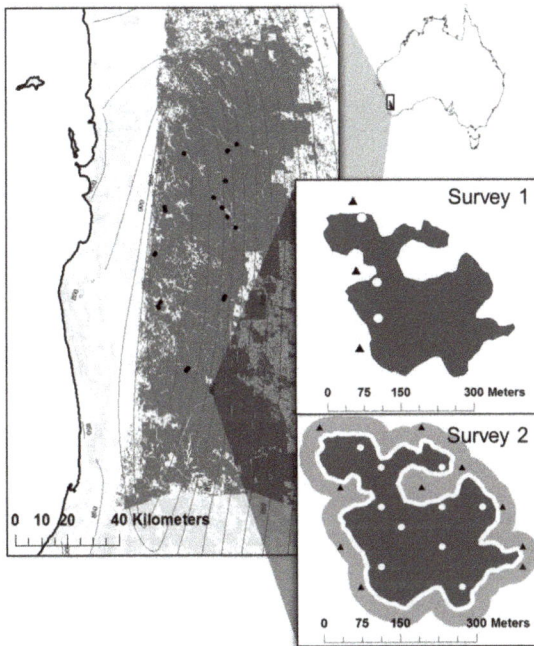

Figure 1. Study site location was the Northern Jarrah Forest of Southwestern Australia. Study sites (black dots) were located in the remaining intact, native vegetation of the forest (dark grey color). Within each study site (inset), three and ten plot locations were randomly selected for severely- (white dots) and minimally-affected (black triangles) forest area for Survey 1 and Survey 2, respectively. The discrete nature of the damage allowed for sampling both severely and minimally drought-affected forest in close proximity. A 10 m forest area buffered severely from minimally-affected areas.

For the second survey in May 2012, a subset of 11 sites was revisited to estimate beetle infestation levels. The life cycle of *P. semipunctata* is known to last up to 9 months [30], so by returning to the sites after this time it ensured nearly all *P. semipunctata* had emerged from infested trees. For this survey, the minimally-affected forest areas were defined as a band of forest 30 m wide surrounding the severely-affected area, buffered by 10 m from the severely-affected area (Figure 1). The distinct and sharp contrast between severely and minimally drought-affected areas allowed for sampling of both populations in close proximity to one another. Twenty plot point locations were chosen randomly on a 40 m × 40 m grid using fGIS within severely and minimally drought-affected forest, and a subset of 6–10 points were used for sampling. On each plot point, the point-centered quarter method of tree sampling was used to select trees [31]. Four trees were selected surrounding each point by measuring the closest tree (*E. marginata* or *C. calophylla*, diameter at breast height over bark (DBHOB) >10 cm) to the plot point in each of four quadrats (NW, NE, SE, SW). The distance from the plot center to the four trees selected was measured and used to determine the density of trees in each forest area based on the methods outlined in Mitchell [31].

2.3. Damage and Tissue Moisture

Either 3 or 4 overstory trees (>10 cm DBHOB, *E. marginata* or *C. calophylla*) were chosen for sampling on each plot. Each sampled tree was assessed using differences in visual condition [32] and placed into one of three crown health classes, based on the severity of discoloration exhibited by crown foliage, adapted from Worrall *et al.* [33]. Trees considered "healthy" had predominantly (>75%) green, turgid foliage and limited evidence of stress, "dying" trees had predominantly yellow, pink, and/or dry foliage, while "recently killed" trees had completely red, dead foliage. On each tree at DBHOB and ground line, four patches of bark, greater than 100 cm^2 in size, were removed from each cardinal direction (8–100 cm^2 patches per tree). Removing the bark exposed the surface of the inner bark/phloem where *P. semipunctata* forms feeding galleries [16]. The area of stem surface damaged by *P. semipunctata* feeding was determined to the nearest 1 cm^2 using a 100 cm^2 transparent grid, placed over each patch of exposed inner bark. In order to estimate the phloem and outer sapwood tissue moisture, which is thought to be important in *P. semipunctata* infestation success [14,34], a 2 cm diameter plug of tissue was collected from each sampled patch using a 19 mm diameter drill bit. Tissue samples were weighed in the field immediately following collection, taken to the laboratory, dried at 70 °C for five days, and re-measured to determine moisture content.

2.4. Infestation Levels

For each tree, the loose outer bark was scraped from the bottom 2 m of the stem to more clearly detect *P. semipunctata* exit holes. Following outer bark removal, the total number of *P. semipunctata* exit holes (defined as being 8–12 mm in diameter) was counted. The density of *P. semipunctata* (exit holes per m^2 stem surface area) was derived for each tree using DBHOB and assuming a perfect cylinder of 2 m in height. These methods closely followed other studies, where Cerambycid beetle populations were estimated using only the bottom 2 m of trees [35], and is justified by within-tree studies that have shown this area represents the majority of beetle emergence [36] Seaton (unpublished data). To calculate adult emergence (ha^{-1}) tree densities found from surveys, and population levels per tree for each forest area (averaged over all sites measured) were then scaled-up and used to calculate total estimated borer populations for each minimally and severely drought-affected forest areas. Sample trees crown health were similarly assigned into classes based on Worrall *et al.* [33], where; "healthy" showed limited evidence of crown dieback from the disturbance (>90% of original crown present), "moderate dieback" included trees showing evidence of partial crown dieback, while retaining some (>10%) original foliage, "severe dieback" included trees which lost all original foliage (and had coppice and epicormic shoots present) and "recently killed" trees had lost all foliage during the disturbance and either failed to resprout or all resprouts were dead.

2.5. Data Analysis

2.5.1. Damage and Tissue Moisture

Since stem damage and tissue moisture were found to be similar between *E. marginata* and *C. calophylla* following Mann–Whitney tests (damage $U = 2.39$, df = 1, $p = 0.1217$, moisture $U = 0.81$, df = 1, $p = 0.3679$), they were grouped for all analyses. To examine the association between *P. semipunctata* damage and crown health class, the proportion of inner bark damaged (severity) by *P. semipunctata* and tissue moisture were analyzed using nonparametric Kruskal–Wallis tests followed by Mann–Whitney tests with Bonferonni correction in SAS (Statistical Analysis System, SAS Institute, version 9.3, Cary, NC, USA) due to non-normally distributed data. Means ($\pm$SE) are shown as appropriate. To determine whether inner bark damage (%) is related to tissue moisture (%), Spearman Rank correlation procedure (PROC CORR) was used in SAS.

2.5.2. Infestation Levels

We compared *E. marginata* and *C. calophylla* infestation levels using Mann–Whitney tests and were found to be not significantly ($U = 1.07$, df = 1, $p = 0.285$) different. Data were then pooled for all analysis. To compare infestation levels of *P. semipunctata* in severely and minimally drought-affected forest areas and the percentage of trees and infestations of *P. semipunctata* among each crown health class in the severely-affected forest area were analyzed using nonparametric Kruskal–Wallis tests followed by Mann–Whitney tests with Bonferonni correction due to unequal variances among health classes in Genstat v16 for Windows, (VSN International, Hempstead, United Kingdom). Means ($\pm$SE) are shown as appropriate.

3. Results

3.1. Damage and Tissue Moisture

Phoracantha semipunctata damage to the inner bark/phloem was found to vary with crown condition ($H_2 = 37.97$, df = 2, $P < 0.0001$), with the greatest damage in trees considered to be recently-killed when surveyed (Figure 2). There was a significant negative relationship ($r = -0.4009$, $p < 0.0001$) between tissue moisture and the proportion of inner bark damaged (Figure 3).

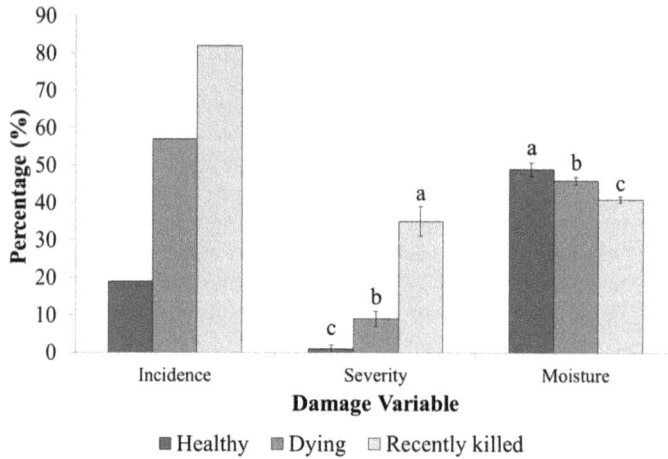

Figure 2. The pooled percentage incidence of *Phoracantha semipunctata* damage, the average severity of damage (% inner bark surface area damaged) and mean tissue moisture content (%) in healthy, dying and recently killed trees across 20 sites in the Northern Jarrah Forest, southwestern Australia (*n* = 60). Significant differences among crown health classes were found for damage severity and tissue moisture from Kruskal–Wallis tests at alpha = 0.05. Columns sharing the same letter within a variable are not different from individual Mann–Whitney tests with Bonferroni correction. Error bars represent the standard error of the mean.

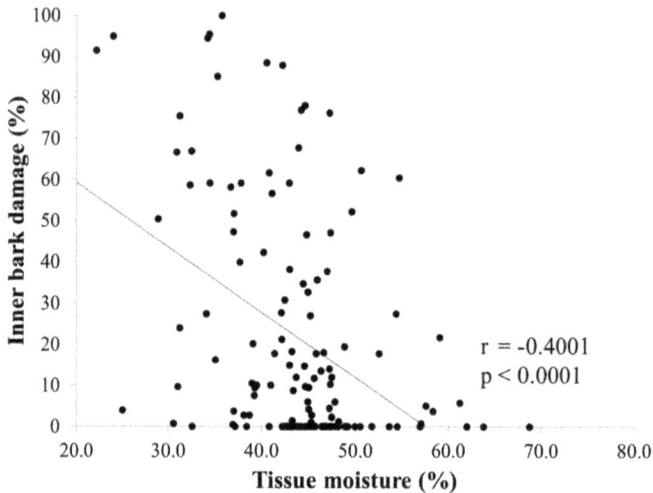

Figure 3. Correlation between the percentage tissue moisture and the percentage of inner bark damaged from *Phoracantha semipunctata* in the Northern Jarrah Forest, southwestern Australia. A significant relationship was determined from Spearman Rank correlation at alpha = 0.05.

3.2. Infestation Levels

On average in the severely-affected forest area, *P. semipunctata* infestation levels (measured as exit holes) were 80 times higher than in the minimally-affected area (3.44 ± 0.39 m^{-2} *vs.* 0.05 ± 0.04 m^{-2}, $H = 60.85$, df = 1, $P < 0.001$) (Figure 4). This scaled-up to an estimated 2265 beetles per hectare for a severely-affected forest and 41 beetles per hectare for a minimally-affected forest, when considering only the lower 2 m of trees.

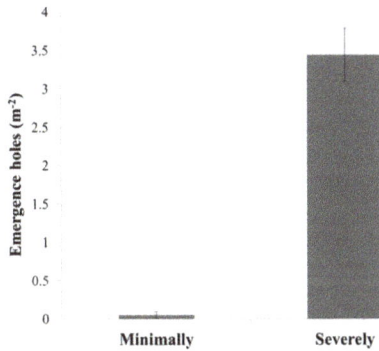

Figure 4. Mean density of *Phoracantha semipunctata* emergence holes (m^{-2}) per bottom 2 m of stems in minimally and severely drought-affected forest areas in Northern Jarrah Forest, southwestern Australia. Significant differences between forest areas were found for emergence holes from Kruskal–Wallis tests at alpha = 0.05. Error bars represent the standard error of the mean, n = 92,304.

In the severely-affected forest, the majority of trees showed severe dieback (>50%) (H = 21.33, df = 3, P < 0.001) with the highest incidence of trees being infested by *P. semipunctata* (Figure 5).

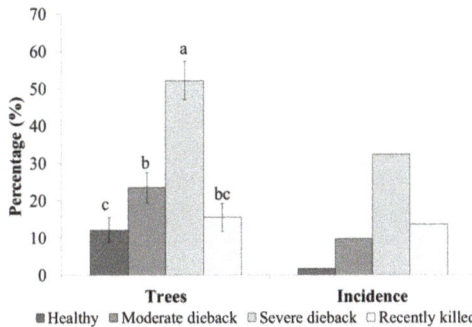

Figure 5. The mean percentage of trees and the pooled percentage incidence of *Phoracantha semipunctata* emergence holes in trees from four crown health classes within the severely drought-affected forest area across 11 sites in the Northern Jarrah Forest, southwestern Australia. Significant differences among trees with different health classes were found from Kruskal–Wallis tests at alpha = 0.05. Columns sharing the same letter for percentage trees are not different from Mann–Whitney tests with Bonferroni correction. Error bars represent the standard error of the mean.

Density of *P. semipunctata* emergence holes varied with crown health class (H = 21.26, df = 3, P < 0.001), where infestation levels increased as crown health deteriorated (Figure 6). Emergence holes in trees with severe dieback and recently killed were 17 times higher compared to trees that were healthy or had moderate dieback (Figure 6). Density of *P. semipunctata* emergence holes was found to be different between sites (H = 37.22, df = 10, P < 0.001).

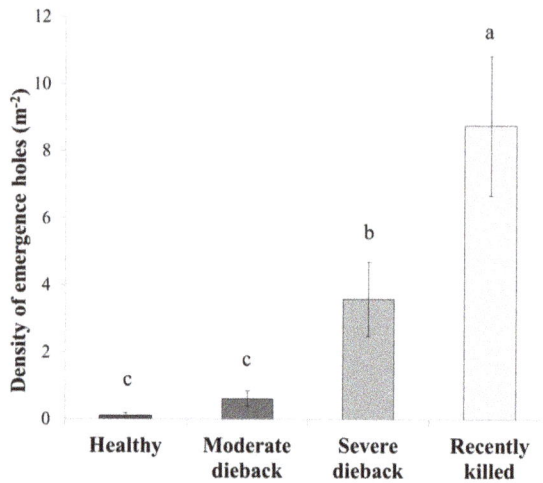

Figure 6. The density of emergence holes (m^{-2}) of *Phoracantha semipunctata* in trees with different crown health classes within severely-affected forest areas across 11 sites in the Northern Jarrah Forest, southwestern Australia. Significant differences between emergence holes in trees with health classes were found from Kruskal–Wallis tests at alpha = 0.05. Columns sharing the same letter are not different from individual Mann–Whitney tests with Bonferroni correction. Error bars represent the standard error of the mean.

4. Discussion

This is the first study to report that, following a severe drought, canopy collapse was one of the main contributors to outbreaks of *P. semipunctata*, in a native ecosystem in Australia. While this interaction is well known in California and other regions where *P. semipunctata* and its hosts have been introduced, it has been rarely reported from native ecosystems [19]. These findings are significant since the tree species in the Northern Jarrah Forest have historically been resistant to the combined effects of drought and forest pests [37]. This beetle outbreak, along with other pest outbreaks in similar ecosystems [23], suggest that climate shifts are changing host-pest interactions in the region.

Host defense against forest pests is expected to be altered with climate change and associated increases in host stress [12]. Inner bark tissue moisture has been found to be an important defense against the successful development of *P. semipunctata* in eucalypts [14,34]. Sufficient inner bark and phloem moisture is thought to force larvae to feed in poorer quality outer bark tissue, which can reduce beetle growth and survival and limit damage to the host [14]. If beetles are unable to avoid the moist inner bark, death of larvae can occur from excessive moisture, or drowning [34]. The association between low tissue moisture and high inner bark damage reported here, during severe drought, provides some support for previous findings [14,34]. However, other related factors, including the concentration of non-structural carbohydrates in tissues, reduced secondary defense compounds, and elevated temperatures may have also contributed to larval development [38–40]. Further research is necessary to determine the environmental changes that occur in hosts during periods of drought stress, and the relative influence of these factors on *P. semipunctata* establishment and success in the Northern Jarrah Forest.

This is the first study to estimate population levels of *P. semipunctata* in the Northern Jarrah Forest, and indeed any natural forest in its native range. Infestation levels of *P. semipunctata* observed in minimally-affected forest are likely to be near, background levels for the Northern Jarrah Forest, since the level of drought stress observed in these areas is consistent with that observed during the seasonal summer drought. In comparison, background levels of other endemic Cerambycids, such

Forests **2015**, *6*, 3868–3881

as *E. rufulus* are generally higher [13]. However, outbreak population levels of *P. semipunctata* were found to be comparable to levels of *E. rufulus* in oak (*Quercus* spp.) trees [35]. The implications of high populations of *P. semipunctata* are not well-known in its native ecosystems; however, this study clearly shows that beetles stayed confined to severely-stressed areas, unlike the pine bark beetles in North America's forests which attacked healthy trees during outbreaks [41]. This suggests that populations of *P. semipunctata* fall considerably after the cessation of drought. Despite this, with continued drying in southwestern Australia, the combination of drought and heat may lead to more frequent and severe outbreaks of *P. semipunctata*.

The drought-induced canopy collapse in the Northern Jarrah Forest was not unlike other mortality events observed in recent years [42], in terms of its severity and extent. While results from this study strongly suggests *P. semipunctata* populations are likely to rise in the coming years with increases in forest drought stress, the effects of increased beetle populations on the Northern Jarrah Forest is unclear. If the behavior of *P. semipunctata* in the Northern Jarrah Forest closely follows that of other endemic Cerambycids [35,43], population outbreaks such as the one observed following the 2011 collapse may predispose trees to suffer higher mortality during less severe drought episodes in future years. Increased population monitoring may enable us to determine the magnitude of forest changes from *P. semipunctata* outbreaks in relation to climate warming.

5. Conclusions

Drought-affected trees in this study were vulnerable to beetle attack from the native Cerambycid, *P. semipunctata*, with trees killed by the 2011 drought containing the highest infestation levels. Low levels of tissue moisture in dead and dying trees, which is known to make other hosts susceptible to *P. semipunctata*, was correlated with high levels of beetle damage. Since *P. semipunctata* was largely restricted to trees severely-affected by drought, and these trees were concentrated in discrete areas, there is minimal evidence that *P. semipunctata* can attack and kill trees not experiencing severe water stress. However, as a result of climate change, drought prone areas are expected to become more common and could result in more frequent outbreaks.

Acknowledgments: The authors thank Department of Parks and Wildlife for their support in accessing the Northern Jarrah Forest. Financial support for this research was provided in part through a National Climate Change Adaptation Research Facility (NCCARF) grant and Murdoch University.

Author Contributions: Stephen Seaton designed the original study, undertook fieldwork and running of the experiment, analyzed data and co-wrote the manuscript. George Matusick assisted with design and fieldwork, analyzed data and co-wrote the manuscript. Katinka Ruthrof assisted with design and fieldwork and co-wrote the manuscript. Giles E. St. J. Hardy assisted with designing the experiment and carried out manuscript editing.

Conflicts of Interest: The authors declare no conflict of interest.

References

1. Negrón, J.F.; McMillin, J.D.; Anhold, J.A.; Coulson, D. Bark beetle-caused mortality in a drought-affected ponderosa pine landscape in Arizona, USA. *For. Ecol. Manag.* **2009**, *257*, 1353–1362. [CrossRef]

2. Hart, C.J. Drought induces spruce beetle outbreaks across northwestern Colorado. *Integr. Pest Manag. Rev.* **2014**, *6*, 247–252.

3. Knight, S.K.; Brown, J.P.; Long, R.P. Factors affecting the survival of ash (*Fraxinus spp.*) trees infested by emerald ash borer (*agrilus planipennis*). *Biol. Invasions* **2013**, *15*, 371–383. [CrossRef]

4. Intergovernmental Panel on Climate Change. *Climate Change 2014: The Physical Science Basis. Contribution of Working Group I to the Fifth Assessment Report of the Intergovernmental Panel on Climate Change*; Cambridge University Press: New York, NY, USA, 2014.

5. Bentz, B.J.; Régnière, J.; Fettig, C.J.; Hansen, E.M.; Hayes, J.L.; Hicke, J.A.; Kelsey, R.G.; Negrón, J.F.; Seybold, S.J. Climate change and bark beetles of the western United States and Canada: Direct and indirect effects. *BioScience* **2010**, *60*, 602–613. [CrossRef]

6. Diffenbaugh, N.S.; Pal, J.S.; Giorgi, F.; Gao, X. Heat stress intensification in the Mediterranean climate change hotspot. *Geophys. Res. Lett.* **2007**, *34*, L11706. [CrossRef]

Forests **2015**, *6*, 3868–3881

7. Bates, B.C.; Hope, P.; Ryan, B.; Smith, I.; Charles, S. Key findings from the indian ocean climate initiative and their impact on policy development in Australia. *Clim. Chang.* **2008**, *89*, 339–354. [CrossRef]

8. Peñuelas, J.; Lloret, F.; Montoya, R. Severe drought effects on Mediterranean woody flora in Spain. *For. Sci.* **2001**, *47*, 214–219.

9. Coulson, R.N.; Stephen, F.M. Impacts of insects in forest landscapes-implications for forest health management. In *Invasive Forest Insects, Introduced Forest Trees, and Altered Ecosystems*; Paine, T.D., Ed.; Springer: Dordrecht, The Netherlands, 2006; pp. 101–125.

10. Hebertson, E.G.; Jenkins, M.J. Climate factors associated with historic spruce beetle (coleoptera: Curculionidae) outbreaks in Utah and Colorado. *Environ. Entomol.* **2008**, *37*, 281–292. [CrossRef] [PubMed]

11. Boucher, T.V.; Mead, B.R. Vegetation change and forest regeneration on the Kenai Peninsula, Alaska following a spruce beetle outbreak, 1987–2000. *For. Ecol. Manag.* **2006**, *227*, 233–246. [CrossRef]

12. Flower, C.E.; Gonzalez-Meler, M.A. Responses of temperate forest productivity to insect and pathogen disturbances. *Annu. Rev. Plant Biol.* **2015**, *66*, 547–569. [CrossRef] [PubMed]

13. Stephen, F.M.; Salisbury, V.B.; Oliveira, F.L. Red oak borer, enaphalodes rufulus (coleoptera: Cerambycidae), in the Ozark mountains of Arkansas, USA: An unexpected and remarkable forest disturbance. *Integr. Pest Manag. Rev.* **2001**, *6*, 247–252. [CrossRef]

14. Hanks, L.M.; Paine, T.D.; Millar, J.G.; Campbell, C.D.; Schuch, U.K. Water relations of host trees and resistance to the phloem-boring beetle *Phoracantha semipunctata* F. (Coleoptera: Cerambycidae). *Oecologia* **1999**, *119*, 400–407. [CrossRef]

15. Duffy, E.A.J. *A Monograph of the Immature Stages of Australasian Timber Beetles (Cerambycidae)*; British Museum (Natural History), Ed.; Trustees of the British Museum: London, UK, 1963; p. 235.

16. Hanks, L.M.; McElfresh, J.S.; Millar, J.G.; Paine, T.D. *Phoracantha semipunctata* (coleoptera: Cerambycidae), a serious pest of eucalytpus in California: Biology and laboratory-rearing procedures. *Ann. Entomol. Soc. Am.* **1993**, *86*, 95–102. [CrossRef]

17. Scriven, G.T.; Reeves, E.L.; Luck, R.F. Beetle from Australia threatens eucalyptus. *Calif. Agric.* **1986**, *40*, 4–6.

18. Wang, Q. A taxonomic revision of the Australian genus phoracantha newman (coleoptera: Cerambycidae). *Invertebr. Taxon.* **1995**, *9*, 865–958. [CrossRef]

19. Curry, S.J. The association of insects with eucalypt dieback in south western Australia. In *Eucalypt Dieback in Forests and Woodlands*; Old, K.M., Kile, G.A., Eds.; CSIRO: Melbourne, Australia, 1981; pp. 130–133.

20. Clarke, J. The marri borer (*tryphocaria hamata*). *J. Agric. West. Aust.* **1925**, *2*, 513–517.

21. Hooper, R.J.; Sivasithamparam, K. Characterization of damage and biotic factors associated with the decline of eucalyptus wandoo in southwest western Australia. *Can. J. For. Res.* **2005**, *35*, 2589–2602. [CrossRef]

22. Matusick, G.; Ruthrof, K.; Hardy, G. Drought and heat triggers sudden and severe dieback in a dominant Mediterranean-type woodland species. *Open J. For.* **2012**, *2*, 183–168. [CrossRef]

23. Hooper, R.J.; Wills, A.; Shearer, B.L.; Sivasithamparam, K. A redescription and notes on biology of Cisseis fascigera obenberger (Coleoptera: Buprestidae) on declining eucalyptus wandoo in south-western Australia. *Aust. J. Entomol.* **2010**, *49*, 234–244. [CrossRef]

24. Matusick, G.; Ruthrof, K.X.; Brouwers, N.C.; Dell, B.; Hardy, G.S.J. Sudden forest canopy collapse corresponding with extreme drought and heat in a Mediterranean-type eucalypt forest in southwestern Australia. *Eur. J. For. Res.* **2013**, *132*, 497–510. [CrossRef]

25. Hanks, L.M.; Paine, T.D.; Millar, J.G. Influence of the larval environment on performance and adult body size of the wood-boring beetle *Phoracantha semipunctata*. *Entomol. Exp. Appl.* **2005**, *114*, 25–34. [CrossRef]

26. Gentilli, J. Climate of the jarrah forest. In *The Jarrah Forest: A Complex Mediterranean Ecosystem*; Dell, B., Havel, J., Eds.; Kluwer Academic Publ: Dordrecht, The Netherlands, 1989; pp. 23–40.

27. Marchant, N.G. Species diversity in the southwestern flora. *J. R. Soc. West. Aust.* **1973**, *56*, 23.

28. McArthur, W.M.; Churchwood, H.M.; Hick, P.T. *Landform and Soils of the Murray River Catchment Area of Western Australia*, 3rd ed.; Management, D.O.L.R., Ed.; Commonwealth Scientific and Industrial Research Organization, Division of Land Resources Management: Melbourne, Australia, 1977; pp. 1–23.

29. Brouwers, N.; Matusick, G.; Ruthrof, K.; Lyons, T.; Hardy, G. Landscape-scale assessment of tree crown dieback following extreme drought and heat in a Mediterranean eucalypt forest ecosystem. *Land. Ecol.* **2013**, *28*, 69–80. [CrossRef]

30. Paine, T.D.; Dreistadt, S.H.; Millar, J.G. *Eucalyptus Longhorned Borers*; UC Statewide Integrated Pest Management Program; University of California: Davis, CA, USA, 2009.

31. Mitchell, K. *Quantitative Analysis by the Point-Centered Quarter Method*; Department of Mathematics and Computer Science: Geneva, Switzerland, 2007; pp. 1–37.

32. Flower, C.E.; Knight, K.S.; Rebbeck, J.; Gonzalez-Meler, M.A. The relationship between emerald ash borer (*agrilus planipennis fairmaire*) and ash (*fraxinus spp.*) tree decline: Using visual ash condition assessments and leaf isotope measurements to assess pest damage. *For. Ecol. Manag.* **2013**, *303*, 143–147. [CrossRef]

33. Worrall, J.J.; Egeland, L.; Eager, T.; Mask, R.A.; Johnson, E.W.; Kemp, P.A.; Shepperd, W.D. Rapid mortality of populus tremuloides in southwestern Colorado, USA. *For. Ecol. Manag.* **2008**, *255*, 686–696. [CrossRef]

34. Hanks, L.M.; Paine, T.D.; Millar, J.G. Mechanisms of resistance in eucalyptus against larvae of the eucalyptus longhorned borer (coleoptera: Cerambycidae). *Environ. Entomol.* **1991**, *20*, 1583–1588. [CrossRef]

35. Fierke, M.K.; Kelley, M.B.; Stephen, F.M. Site and stand variables influencing red oak borer, enaphalodes rufulus (coleoptera: Cerambycidae), population densities and tree mortality. *For. Ecol. Manag.* **2007**, *247*, 227–236. [CrossRef]

36. Timms, L.L.; Smith, S.M.; de Groot, P. Patterns in the within-tree distribution of the emerald ash borer (*agrilus planipennis fairmaire*) in young, green-ash plantations of south-western Ontario, Canada. *Agric. For. Entomol.* **2008**, *8*, 313–321. [CrossRef]

37. Abott, I.; Loneragan, O. *Ecology of Jarrah (Eucalyptus Marginata) in the Northern Jarrah Forest of Western Australia*; Department of Conservation and Land Management: Perth, Australia, 1986.

38. Huberty, A.F.; Denno, R.F. Plant water stress and its consequences for herbivorous insects: A new synthesis. *Ecology* **2004**, *85*, 1383–1398. [CrossRef]

39. Salle, A.; Nageleisen, L.-M.; Lieutier, F. Bark and wood boring insects involved in oak declines in Europe: Current knowledge and future prospects in a context of climate change. *For. Ecol. Manag.* **2014**, *328*, 79–93. [CrossRef]

40. Koricheva, J.; Larsson, S.; Haukioja, E. Insect performance on experimentally stressed woody plants: A meta-analysis. *Ann. Rev. Entomol.* **1998**, *43*, 195–216. [CrossRef] [PubMed]

41. Raffa, K.F.; Aukema, B.H.; Bentz, B.J.; Carroll, A.L.; Hicke, J.A.; Turner, M.G.; Romme, W.H. Cross-scale drivers of natural disturbances prone to anthropogenic amplification: The dynmaics of bark beetle eruptions. *BioScience* **2008**, *58*, 501–517. [CrossRef]

42. Allen, C.D.; Macalady, A.K.; Chenchouni, H.; Bachelet, D.; McDowell, N.; Vennetier, M.; Kitzberger, T.; Rigling, A.; Breshears, D.D.; Hogg, E.H.; *et al.* A global overview of drought and heat-induced tree mortality reveals emerging climate change risks for forests. *For. Ecol. Manag.* **2010**, *259*, 660–684. [CrossRef]

43. Fan, Z.; Kabrick, J.M.; Spetich, M.A.; Shifley, S.R.; Jensen, R.G. Oak mortality associated with crown dieback and oak borer attack in the ozark highlands. *For. Ecol. Manag.* **2008**, *255*, 2297–2305. [CrossRef]

forests

MDPI

Article

Species Distribution Model for Management of an Invasive Vine in Forestlands of Eastern Texas

Hsiao-Hsuan Wang [1,*], Tomasz E. Koralewski [2], Erin K. McGrew [1], William E. Grant [1] and Thomas D. Byram [2,3]

[1] Department of Wildlife and Fisheries Sciences, Texas A&M University, College Station, TX 77843, USA; ekmcgrew@email.tamu.edu (E.K.M.); wegrant@tamu.edu (W.E.G.)

[2] Department of Ecosystem Science and Management, Texas A&M University, College Station, TX 77843, USA; tkoral@tamu.edu (T.E.K.); t-byram@tamu.edu (T.D.B.)

[3] Texas A&M Forest Service, College Station, TX 77843, USA

* Correspondence: hsuan006@tamu.edu; Tel.: +1-979-845-5702; Fax: +1-979-845-3786

Academic Editor: Diana F. Tomback

Received: 14 September 2015; Accepted: 18 November 2015; Published: 27 November 2015

Abstract: Invasive plants decrease biodiversity, modify vegetation structure, and inhibit growth and reproduction of native species. Japanese honeysuckle (*Lonicera japonica* Thunb.) is the most prevalent invasive vine in the forestlands of eastern Texas. Hence, we aimed to identify potential factors influencing the distribution of the species, quantify the relative importance of each factor, and test possible management strategies. We analyzed an extensive dataset collected as part of the Forest Inventory and Analysis Program of the United States Department of Agriculture (USDA) Forest Service to quantify the range expansion of Japanese honeysuckle in the forestlands of eastern Texas from 2006 to 2011. We then identified potential factors influencing the likelihood of presence of Japanese honeysuckle using boosted regression trees. Our results indicated that the presence of Japanese honeysuckle on sampled plots almost doubled during this period (from 352 to 616 plots), spreading extensively, geographically. The probability of invasion was correlated with variables representing landscape conditions, climatic conditions, forest features, disturbance factors, and forest management activities. Habitats most at risk to invasion under current conditions occurred primarily in northeastern Texas, with a few invasion hotspots in the south. Estimated probabilities of invasion were reduced most by artificial site regeneration, with habitats most at risk again occurring primarily in northeastern Texas.

Keywords: biodiversity; biological invasions; boosted regression trees; Japanese honeysuckle; likelihood of invasion; *Lonicera japonica* Thunb.

1. Introduction

Invasive species have caused dramatic economic damage and loss of biodiversity worldwide and non-native species continue to invade forest ecosystems in the United States at an accelerating rate [1]. Despite general consensus on the risks posed by alien species, the exact economic and ecological impact of invasions is an active area of debate and, in some cases, certain potential benefits are pointed out, depending on particular species, scenario, and perspective (reviewed in [2]). Nevertheless, there are an estimated 50,000 non-native species on U.S. soil [3]. According to Wilcove *et al.* [4], 57% of the 1055 imperiled plant species in the U.S., for which they managed to collect threat data, were negatively impacted by alien species. Based on the same data, Gurevitch and Padilla [5], estimated that out of 602 plant species in the U.S. imperiled by invaders, 410 (about 68%) could be linked to competition and indirect habitat effects caused by alien plants, but highlighted that invasive plants often act in parallel with other factors, such as habitat loss. Invasive plants reduce forest productivity [6] and degrade forest health [2], and pose a particularly severe threat to the economically and ecologically important forestlands in the southeastern U.S. Southeastern forests contain about 201 million acres of timberland, corresponding to about 58% of U.S. timber production and about 16% of timber production world-wide [7], and also support a high level of biodiversity, with over 3000 native plant species [8]. However, invasive plants threaten both the productivity of timber species essential to economic sustainability of the region [9] and the biodiversity equilibrium essential to the provision of ecosystem services, such as erosion control [10].

Japanese honeysuckle (*Lonicera japonica* Thunb.), which was introduced into the U.S. from China and Japan in the early 1800s [11], is one of the most aggressive invasive species in North America. It has become naturalized in 45 states of the U.S., and the Canadian province of Ontario [11,12]. Its attractive and fragrant flowers made it a valued cultivar, but its long flowering period and hardy growth characteristics allowed it to escape cultivation [11], probably during the 1890s, and spread to most of its present range within about 30 years [13]. Japanese honeysuckle prefers moist woodlands, often growing in tangled clumps along slopes, but is able to tolerate most forest conditions [14]. It grows rapidly, flowers early and for long durations [15], reproduces via animal-dispersed seeds or vegetatively [16], possesses phenotypic plasticity which reduces herbivore damage [17], and has no known predators or pathogens within its invaded range [11]. It negatively affects native vegetation [18], modifying vegetation structure [19], decreasing biodiversity [20,21], and inhibiting natural regeneration of native species, including loblolly pine [22], a predominant tree species in both artificially and naturally regenerated forests of the southeast. While natural pine forest range has been gradually declining, the acreage of pine plantations has been steadily expanding since 1950s, reaching 32 million acres in 1999 with further expected increase up to 54 million acres in 2040 [23]. The expanding plantations and increased timber yield additionally highlight economic potential of the southeastern forests. Approximately 84% of the seedlings planted annually throughout the South are loblolly pine [24]. In the forestlands of eastern Texas, largely corresponding to the western part of the Piney Woods ecoregion, where loblolly pine is of both commercial and ecological importance, Japanese honeysuckle is the most prevalent invasive vine [25].

A recent study identified areas vulnerable to invasion by Japanese honeysuckle throughout Mississippi, stretching northward across western Tennessee and western Kentucky, westward across southern Arkansas, eastward across north-central Alabama, and also in several counties scattered within Virginia [26]. These authors used logistic regression to correlate land characteristics and climatic conditions with presence/absence of Japanese honeysuckle based primarily on analysis of data collected prior to 2006.

In the present study, we calculated the recent range expansion of Japanese honeysuckle within the forestlands of eastern Texas based on analyses of an extensive set of field data collected by the U.S. Forest Service on fixed plots during the period from 2007 to 2011. We then used boosted regression trees to analyze an extensive data set collected as part of the Forest Inventory and Analysis (FIA) program of the United States Department of Agriculture (USDA) Forest Service to identify areas within the forestlands of eastern Texas that were invaded by Japanese honeysuckle during the FIA inventory period that extended from 2007 to 2011. We quantitatively assessed a suite of landscape features, climatic conditions, forest conditions, forest management activities, and disturbances as potential factors affecting the likelihood of invasion. Based on identification of the potentially most influential factors, we predicted likelihoods of future invasions by Japanese honeysuckle under a common management strategy (artificial regeneration of loblolly pine).

Forests **2015**, *6*, 4374–4390

2. Methods

2.1. Study Area and Data Sources

We focused our investigation on eastern Texas, which is an area of vast timber resources [27] that has been invaded by Japanese honeysuckle [25]. The warm and humid climate in the area is conducive to invasion.

To document range expansion of Japanese honeysuckle in eastern Texas between the years of 2006 and 2011, we extracted data on the presence or absence of Japanese honeysuckle using the Southern Nonnative Invasive Plant data Extraction Tool [25] of the USDA Forest Service. In the early 2000s, as part of the Forest Inventory and Analysis (FIA) Program, the Forest Service's Southern Research Station began an intensive regional monitoring of nonnative plant invasions [28]. The FIA Program is a forest inventory program in which each state inventory is reported every five years in most southeastern states [29]. The basic sampling design consists of a lattice of 4047-m^2 hexagons, with one sample plot located randomly within each hexagon [29,30]. Each sample plot consists of four subplots of radius 7.32 m which form a cluster consisting of a central subplot and three peripheral subplots equidistant from each other arrayed in a circle of radius 36.58 m centered on the central plot. On each subplot, inventory crews estimate percent cover by target invasive species [28], and also record a suite of landscape conditions and forest features, as well as past disturbances and forest management activities [30].

2.2. Quantification of the Range Expansion

At present, two state inventory cycles recording the presence of nonnative plant species have been reported (in 2006 and 2011) for eastern Texas [25]. We documented range expansion by noting the presence (with cover) or absence (without cover) of Japanese honeysuckle on each subplot sampled during each of these two inventories and then mapping and counting the plots in each inventory with Japanese honeysuckle present. We also calculated an index of potential rate of spread based on the distance between each plot in which Japanese honeysuckle was first detected during the second survey and the nearest plot in which Japanese honeysuckle had been detected during the first survey, that is, the distance to the nearest known propagule source.

2.3. Identification of Potential Factors Influencing Likelihood of Invasion

Widely recognized factors influencing the likelihood of invasion of Japanese honeysuckle include landscape conditions, climatic conditions, forest features, disturbance factors, and forest management activities [26,31,32]. Landscape conditions include slope and adjacency to water bodies within 300 m. Climatic conditions include mean annual precipitation, and mean annual minimum and maximum temperature. Forest features include stand age, site productivity, basal area, and natural regeneration (growth of existing trees, natural seeding, or both, resulting in a stand at least 50% stocked with live trees of any size). Disturbance factors include distance between the plot and the nearest road, distances between the plot and the nearest known propagule source, and forest disturbance (such as those caused by animals, disease, fire, insects, and/or wind). Forest management activities such as harvesting, site preparation (including clearing, slash burning, chopping, disking, bedding, or other practices clearly intended to prepare a site for reforestation), and artificial regeneration (planting or direct seeding resulting in a stand in at least 50% stocking with trees of any size). Data on several factors in each of these broad categories are available from the traditional FIA dataset for plots within our study area (Table 1). Drawing on the literature on plant invasions, we selected a set of possible variables for plots within our study area from the PRISM database [33,34] and the FIA dataset (Table 1). We also used the same dataset to compute Shannon's index of tree species diversity, H_s, for each plot [35,36]:

$$H_s = -\sum_{i=1}^{n_s} \frac{B_i}{B} \ln(\frac{B_i}{B})$$

(1)

where B and B_i are the total stand basal area and the basal area of trees of species i, respectively, and n_s is the number of tree species.

Table 1. Descriptions, possible values or units of measure, and means or counts of landscape conditions, forest features, disturbance factors, and forest management activities evaluated as potential factors of site invasion by Japanese honeysuckle in forest plots of eastern Texas.

Variable	Value or Unit of Measure	Mean (Range) for Continuous Data/Count for Categorical Data
Landscape conditions		
Slope	Degree	2.07 (0.00–32.50)
Adjacency to water bodies within 300 m	No	1830
	Yes	471
Climate conditions		
Mean annual precipitation	cm	127.06 (99.06–154.94)
Mean annual minimum temperature	°C	12.43 (10.56–16.67)
Mean annual maximum temperature	°C	24.89 (22.78–26.11)
Forest features		
Stand age	Year	35.76 (1–104)
Site productivity	L1: 0–1.39 $m^3 \cdot ha^{-1} \cdot year^{-1}$	9
	L2: 1.40–3.49	154
	L3: 3.50–5.94	660
	L4: 5.95–8.39	931
	L5: 8.40–11.54	541
	L6: 11.55–15.74	95
	L7: >15.74	4
Basal area	m^2	1.26 (0.00–7.10)
Species diversity	Shannon's species diversity	1.02 (0.00–2.53)
Natural regeneration [a]	No	2348
	Yes	46
Disturbance factors		
Distance to the nearest road	D1: <30 m	169
	D2: 30–91	382
	D3: 92–152	273
	D4: 153–305	492
	D5: 306–805	674
	D6: 806–1609	239
	D7: 1610–4828	62
	D8: 4829–8047	3
	D9: >8047	2
Distance to the nearest known propagule source	m	14,343 (0–75,529)
Forest disturbance [a,b]	No	2294
	Yes	100

Table 1. *Cont.*

Variable	Value or Unit of Measure	Mean (Range) for Continuous Data/Count for Categorical Data
Forest management activities		
Harvesting [a]	No	1969
	Yes	425
Site preparation [a]	No	2309
	Yes	85
Artificial regeneration [a]	No	1802
	Yes	592

[a] Normally within the past five years; [b] A disturbance code of "yes" indicates that at least 25% of the trees in a stand are damaged by animals (from deer/ungulates, rabbits, or a combination of animals), disease, fire (from crown or ground fire, either prescribed or natural), insects, and/or wind (including, but not limited to, damages from hurricanes and tornados).

2.4. Data Analysis

We associated the data on presence or absence of Japanese honeysuckle (SNIPET) with the data on landscape conditions, forest features, disturbance factors, and forest management activities (FIA Data and Tools) using the FIA plot identification numbers. We then conducted our analysis with a process of machine learning that uses boosted regression trees, which combines decision trees and a boosting algorithm with a form of logistic regression [37]. For boosted regression trees, the probability (p) of species occurrence ($y = 1$) at a location with the potential explanatory variables (X) is given by $p(y = 1|X)$ and is modeled via the logit: *logit* $p(y = 1|X) = f(X)$. We fitted our model in R (R Development Core Team 2006 version 2.14.1, R Foundation for Statistical Computing, Vienna, Austria) using the gbm package version 1.5-7 [38]. Our aim was to find the best combination of parameters (learning rate and tree complexity) that achieved the minimum predictive error (minimum error for predictions of independent samples). The learning rate, also known as the shrinkage parameter, determines the contribution of each tree to the growing model, and the tree complexity controls whether interactions are fitted. We determined the optimal model following the recommendations of Elith *et al.* [37] by altering the learning rate and tree complexity (the number of split nodes in a tree) until the predictive deviance was minimized without over-fitting, and by limiting our choice of the final model to those that contained at least 1000 trees. We included randomness into the models to reduce over-fitting and also to improve accuracy and speed of the model selection process [39]. We represented randomness using a "bag fraction" that specifies the proportion of data to be selected at each step [37]. We set the default bag fraction at 0.6, meaning that 60% of the data were drawn at random without replacement from the full training set during each iteration. When we obtained the optimal model, we evaluated model performance using a ten-fold cross-validation procedure with re-substitution. For each cross-validation trial, we randomly selected 60% of the dataset for model fitting and we used the excluded 40% for testing. We then used plot Universal Transverse Mercator (UTM) coordinates, x for distance east and y for distance north of the UTM origin, to test for spatial autocorrelation of the residuals of the model, employing Moran's I index. We calculated the response variance explained, the area under the receiver operator characteristic curve (AUC), and the overall accuracy based on the aggregated cross-validation results. We evaluated the reliability and validity of the optimal model as fair ($0.50 < $ AUC $\leqslant 0.75$), good ($0.75 < $ AUC $\leqslant 0.92$), very good ($0.92 < $ AUC $\leqslant 0.97$), or excellent ($0.97 < $ AUC $\leqslant 1.00$) based on the value of AUC [40]. We then used the gbm library to derive the relative influence of each potential explanatory variable in the optimal model and constructed partial dependence plots and fitted values plots for the most influential variables [37].

2.5. Likelihood of Invasion

We used the optimal model to calculate the probability of presence of Japanese honeysuckle on each of the study plots. We then re-estimated the probabilities of presence assuming that artificial regeneration occurred on 25%, 50%, 75%, or 100% of the plots. We represented these different levels of management activity by changing the value of the corresponding element of X (e.g., artificial regeneration) from "No" to "Yes." To visually display the results, we superimposed the probabilities of presence predicted under each level of management activity on maps of the study area using ArcMap^{TM} 10.2.1 (ESRI, Redlands, CA, USA).

3. Results

3.1. Historical Trends in Range Expansion

As indicated by the FIA records from 2011, Japanese honeysuckle spread extensively throughout eastern Texas during only a five-year period. The range expansion has been particularly dramatic in the northern part of the studied area, but the few, initially sparse, hot-spots in the south have also shown dynamism that seems to match that of the north (Figure 1). The presence of Japanese honeysuckle almost doubled from 352 plots (14.69%) in 2006 to 616 plots (25.7%) in 2011, and the number of counties invaded increased from 23 to 41, leaving only two counties un-invaded. The distance between each plot in which Japanese honeysuckle was first detected during the second survey and the nearest plot in which Japanese honeysuckle had been detected during the first survey ranged from 0 km to 75 km, with a mean (±SE) of 16 km (±0.5 km).

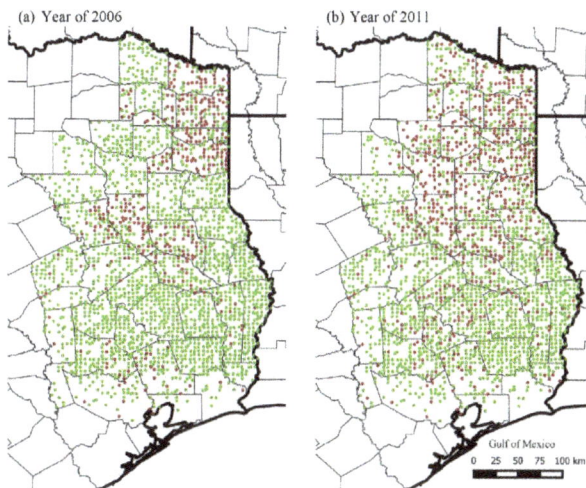

Figure 1. Presence (red dots) and absence (green dots) of Japanese honeysuckle in forested plots sampled in eastern Texas in (**a**) 2006 and (**b**) 2011 as part of the Forest Inventory and Analysis Program of the U.S. Forest Service [25].

3.2. Potential Determinants of Invasion

We explored 250 combinations of tree complexity (ranging from three to seven) and learning rate (ranging from 0.001 to 0.05); these produced models with between 800 and 1400 trees. The optimal model had a tree complexity of five, a learning rate of 0.01, and a total of 1100 trees. Moran's I index ($I = 0.002$) indicated no statistically significant ($p = 0.17$) spatial autocorrelation. Model predictive deviance was 0.894 ± 0.015 with 80.6% of the total response variance explained. The AUC score was

0.855 ± 0.008 ("good" ability to discriminate between species presence and absence) and the overall accuracy was 80.8%. Recursive feature elimination tests showed that five variables could be removed from the model before the resulting predictive deviance exceeded the initial predictive deviance of the model with all variables.

Examination of the relative contribution of the predictor variables indicated that the top five accounted for approximately 65% of the contribution in the overall model (Figure 2). Of the five most influential variables, three were forest features and two were climatic conditions. Basal area was the most influential variable, contributing 14.8%. Mean annual maximum temperature, biodiversity index (Shannon's species diversity), stand age, and precipitation were the second, third, fourth, and fifth most important variables, contributing 13.7%, 12.8%, 12.0%, and 11.7%, respectively. Distances to the nearest known propagule source and distance to the nearest road were selected from disturbance factors. Landscape conditions, climatic conditions, forest features, disturbance factors, and forest management activities had total contributions of 4.0%, 32.1%, 44.8%, 17.3%, and 1.7%, respectively (Figure 2).

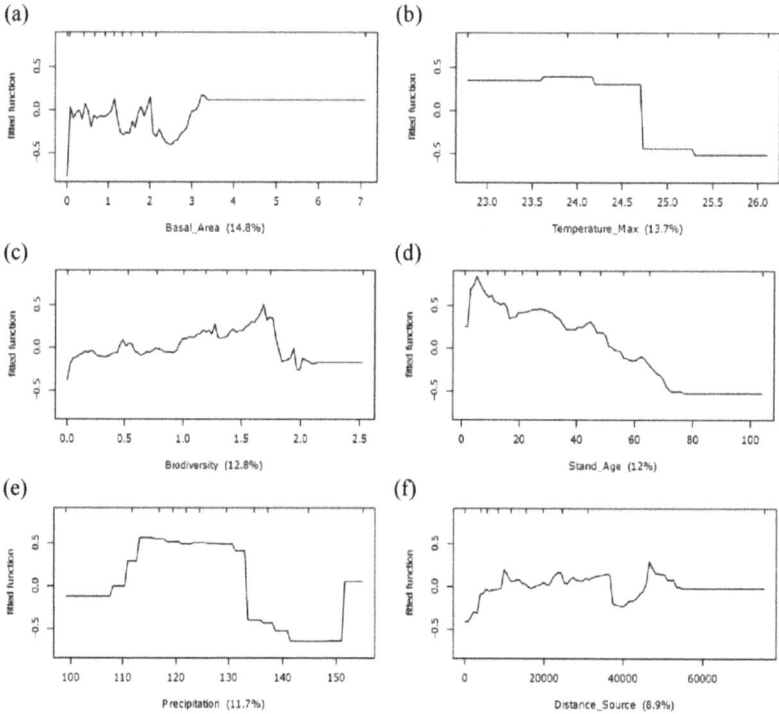

Figure 2. *Cont.*

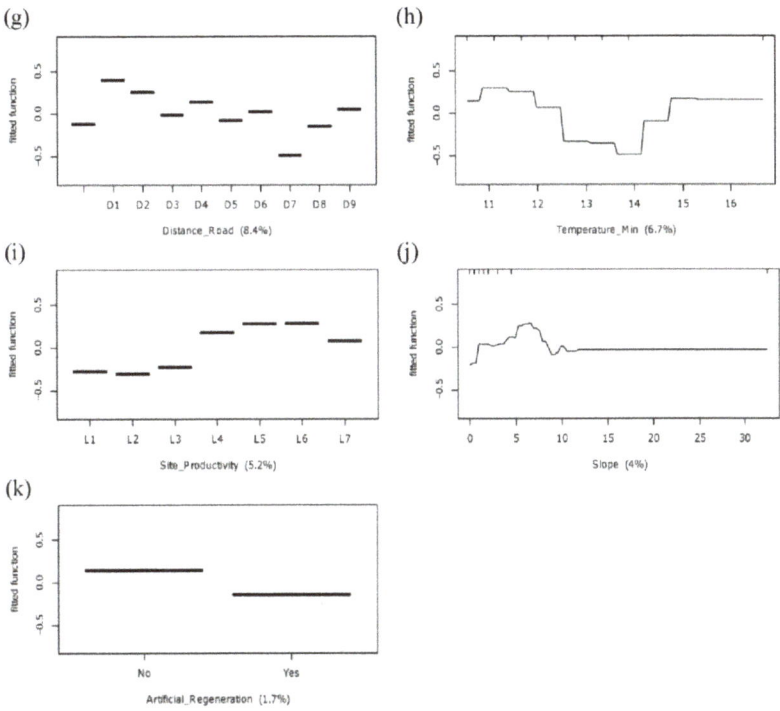

Figure 2. Partial dependence plots for the explanatory variables included in the optimal boosted regression tree models for Japanese honeysuckle based on analyses of the 11 most influential variables. Hash marks at the top of each plot indicate distribution of sample plots along the range of the indicated variable. X axis labels include the relative contributions (%) of each influential variable included in the final model (see Table 1 for the description of variables). Y axes are based on the logit scale used for the indicated variable.

Partial dependence plots indicated that Japanese honeysuckle occurrences were associated with plots with a slope ratio lower than 8% (Figure 2j). Climatic conditions included a mean annual maximum temperature below 24.7 °C (Figure 2b), an annual precipitation between 110 and 133 cm (Figure 2e), and a mean annual minimum temperature between 10.6 °C and 12.5 °C (Figure 2h). Forest features usually included a basal area less than 2 m² (Figure 2a), a biodiversity index between 1.0 and 1.8 (Figure 2c), stand age younger than 50 years (Figure 2d), and site productivity higher than 5.95 m³·ha⁻¹·year⁻¹ (above L3, Figure 2i). Occurrences were more likely in plots with distances to the nearest known propagule less than 35,000 m (Figure 2f) and to the nearest road less than 91 m (below D3, Figure 2g), and with no artificial regeneration within the past five years (Figure 2k). In addition, the fitted values plots (Figure 3) indicated all fitted values in each factor type.

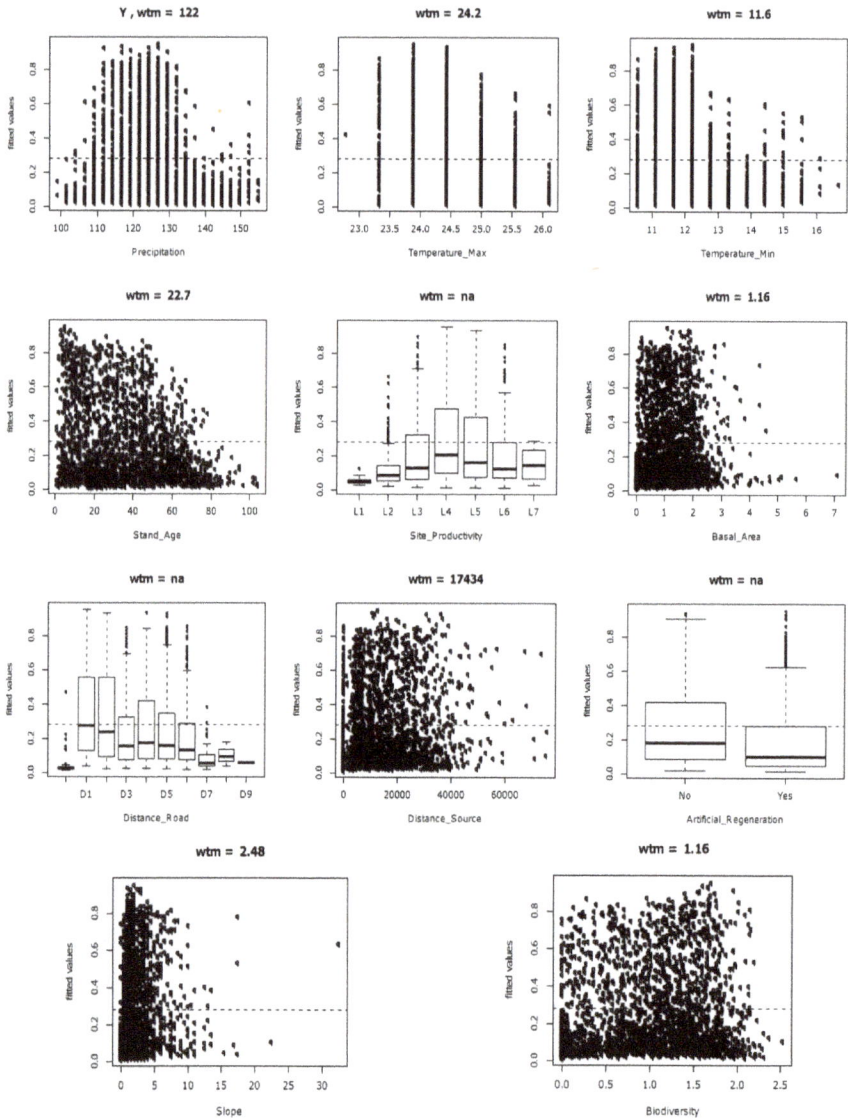

Figure 3. Fitted values plots for the explanatory variables included in the optimal boosted regression tree models for Japanese honeysuckle based on analyses of the 11 most influential variables. Values above each graph indicate the weighted mean of fitted values in relation to each non-factor explanatory variable.

3.3. Likelihood of Invasion under Current Conditions and Different Levels of Management

Estimated probabilities (p) of invasion under current conditions indicated that approximately 70% of the plots fell within the $0 < p \leqslant 0.2$ category, 24% within the $0.2 < p \leqslant 0.4$ category, 5% within the $0.4 < p \leqslant 0.6$ category, and about 1% within the $0.6 < p \leqslant 0.8$ category (Figure 4). The majority of the relatively higher estimated probabilities ($p > 0.4$) were located in northeastern Texas (Figure 5a).

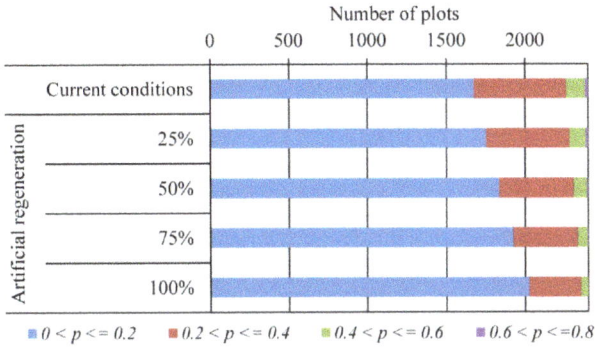

Figure 4. Number of plots within the indicated categories of estimated probability (*p*) of invasion (blue: 0 < *p* ⩽ 0.2, red: 0.2 < *p* ⩽ 0.4, green: 0.4 < *p* ⩽ 0.6, purple: 0.6 < *p* ⩽ 0.8) by Japanese honeysuckle assuming current conditions, and artificial regeneration on 25%, 50%, 75%, or 100% of the plots (chosen at random).

Increasing the level of artificial regeneration from zero to 100% in increments of 25% reduced estimated mean probabilities (*p*) of invasion from 0.17 to 0.16, 0.15, 0.13, and 0.12, respectively, and decreased the number of plots with *p* > 0.2 from 724 to 648, 564, 474, and 373, respectively, (Figure 4). Once again, the majority of the plots with relatively higher estimated probabilities (*p* > 0.2) were located in northeastern Texas (Figure 5b–e).

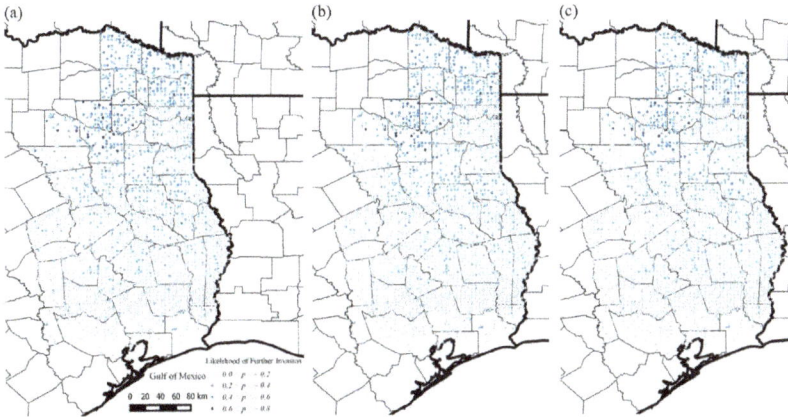

Figure 5. *Cont.*

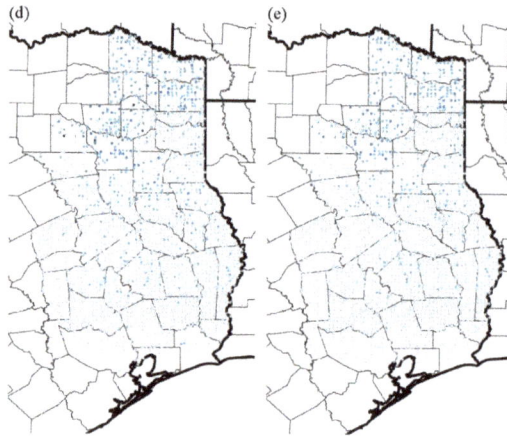

Figure 5. Estimated probabilities of invasion by Japanese honeysuckle in plots sampled in eastern Texas as part of the Forest Inventory and Analysis Program of the USDA Forest Service assuming (**a**) current conditions, (**b**) artificial regeneration on 25%, (**c**) 50%, (**d**) 75%, or (**e**) 100% of the plots (chosen at random).

4. Discussion

The number of sites invaded by Japanese honeysuckle almost doubled within only five years. Clearly, northern sites are far more frequently invaded than those in the south (Figure 1), and the species was identified in, what seems moderate, 25.7% of the sampled sites. However, for a better understanding of the species presence one should perhaps consider that in 2011 less than 5% of the forested counties in eastern Texas remained uninvaded. The species abundance throughout the region seems uneven, but given its ability to spread rapidly, as evidenced by comparison of 2006 *vs.* 2011 presence-absence records, and its current distribution, there is potential for further invasion into other currently uninvaded sites. The change in spatial patterns, as shown by our results, agrees with the earlier work by Wang *et al.* [26] who estimated probabilities of invasion of Japanese honeysuckle throughout the Southern US forestlands based primarily on analysis of data collected during the FIA inventory period that ended in 2006. In particular, they identified areas vulnerable to invasion in eastern Texas, with northeastern sites showing higher probability of invasion. Our index of potential rate of spread, assuming a mean interval of five years between samples on any given plot, suggests a mean potential dispersal velocity of 3200 m/year (16 km/5 years), which is faster than the rates used in the models of Morales and Carlo [41] and Clark *et al.* [42]. Morales and Carlo [41] used a spatially-explicit simulation model to estimate a dispersal velocity of approximately 800 m per year for plants dispersed by frugivores. Clark *et al.* [42] used a stochastic model based on information from a variety of sources to estimate a range of annual velocities of seed dispersal by birds: 90% of the seeds were dispersed 0–100 m from the parent tree, 8% were dispersed 100–500 m, 1.7% were dispersed 500–5000 m and 0.3% were >5000 m. Our index was almost surely biased by exclusion of propagule sources in relatively close, un-sampled areas; the FIA field data were collected from a lattice of 4047-m^2 hexagons, with only one sample plot located randomly within each hexagon [29]. On the other hand, given that seeds of Japanese honeysuckle are dispersed by wide-ranging animals, such as white-tailed deer, a relatively high dispersal velocity is not surprising.

While several factors affect a site's susceptibility to Japanese honeysuckle invasion, forest features and climate conditions seem to play predominant roles. The majority of the species occurrences have been noted on sites with very low basal area, although the exact impact of basal area is somewhat ambiguous considering the interlacing positive and negative effects upon the fitted function along the variable's distribution. A clear and consistent effect, however, may be observed in the case of other major contributors from the forest features group of variables. In particular, the youngest stands are most invasion-prone, and the susceptibility gradually decreases with age. High site productivity appears to favor invasions, as well. Logically, relatively young and highly-productive sites provide favorable conditions for any species, native and invasive ones alike [43]. The main factor identified as a limitation to mature forest stand invasions by Japanese honeysuckle was limited light availability, even though the species can tolerate a relatively low light environment [44,45]. Relatively open habitats facilitate early age flowering and prolific seed production. Halls [46] showed that Japanese honeysuckle in eastern Texas reproduced fruit at three years of age when plants were grown in the open, but at age five when grown in the shade. In addition, signs of stress developed after two years of growth under 8% of ambient light [47]. Even though new growth was initiated each spring, leaders would subsequently die back and a portion of the current leaf crop would abscise following maturation of the flush [47].

Japanese honeysuckle invasions were more abundant on sites with medium-high biodiversity index values. Although it seems counterintuitive that a new species would successfully establish itself in a diverse and likely balanced community (a classic view by Elton; [48]), post-invasion dynamics within this community as well as fate of the invader remain unknown. Debate is ongoing with arguments and evidence both supporting and opposing the classic perspective, however [48–51]. This disagreement also was noted between theoretical analyses and field studies (reviewed in [52]). In a recent case study, Wang and Grant [53] noted that sites with dominant loblolly pine overstory were less diverse than mixed evergreen-deciduous and mostly-deciduous stands. They suggested that the evergreen canopy was the likely factor diminishing risk of encroachment by invasive shrubs due to the decreased amount of light reaching the understory. On the contrary, the mixed canopy partially opens up from fall to early spring allowing for more light to reach the understory during that period. Indeed, the interspecies feedback loops [54,55] that characterize ecological communities challenge us to focus on the processes by which resident species affect the availability of physical resources, such as light, with which the likelihood of species invasions often are correlated.

Apart from favorable forest features, climate conditions of eastern Texas forest stands also appear to facilitate Japanese honeysuckle invasions. Leatherman [15] reported that Japanese honeysuckle in Eastern North America expands generally south of an isotherm where mean January temperature is -1 °C, north of an isotherm where only 5% of January daily low temperatures are <0 °C, and east of a mean annual precipitation of 1016 mm. Although Japanese honeysuckle may tolerate cool winter temperatures and regrow vigorously the following summer, late spring frosts that damage new growth and short growing seasons limit the northern distribution [13,56]. Sasek and Strain [57] suggested that projected future climate change may allow Japanese honeysuckle to expand its northern range; however, its southern distribution may be limited by mild winter temperatures that are insufficient for seed stratification. Our simulated intensity of invasion generally decreased with increasing mean annual temperature and decreasing annual precipitation.

Among the investigated disturbance factors, proximities of the nearest known propagule source and road were significantly correlated with the distribution of Japanese honeysuckle. Propagule pressure has been shown to affect the distribution and abundance of many invasive plants [58,59] and, like many other invasive plants, recruitment limitation tends to slow the range expansion of Japanese honeysuckle [60,61]. These results support the concept that invasions are best mitigated during the early stages of recruitment and establishment [62,63] or via control of outlier populations [64]. The control of biological invasions has become a topic discussed at the highest scientific, policy and management levels [65]. While long-term invasive plant management strategies are recognized as essential to reduce the ecological damage and economic costs associated with invasions, the development and implementation of such plans lags considerably behind [66]. Most of invasive plant management strategies, including that of Japanese honeysuckle, have emphasized controlling highly infested areas, despite this knowledge.

Roads have been shown as invasion corridors that facilitate the spread of invasive plants [67,68]. This is especially likely for an invasive vine such as Japanese honeysuckle, capable of plastic responses to heterogeneous habitat and thus of rapid growth in light gaps such as road-sides [69]. If there are no structures available for twining, Japanese honeysuckle can grow in a mat form along the ground, increasing the likelihood that roads could act as corridors for, rather than barriers to, spread. Spread of the species along roads may be additionally augmented by seed dispersal, also across the road, by wildlife. Japanese honeysuckle has been recognized as an important browse and forage species, especially for white-tailed deer, but also for other wildlife [70–72] throughout much of the eastern and southern United States, particularly during poor mast years and during winter when other food sources are scarce or inaccessible [73,74]. For example, in areas of northern Alabama managed primarily for loblolly pine production, Japanese honeysuckle constituted 49.4% of the year-round diet of white-tailed deer, whereas no other single food item accounted for more than 6% [75]. White-tailed deer can move over large distances [76–78] and have been observed utilizing roadways [79,80] and, thus, may be important vectors of Japanese honeysuckle seed dispersal.

The roles of other studied factors, namely landscape conditions and forest management activities, were rather minor. About 75% of the plots in eastern Texas currently occupied by Japanese honeysuckle occur at the northern part of the area, and about 98% occur on lands with a slope ratio less than 8%, where favorable habitats are created on well-drained sites [81]. In general, the species is noticeably absent on coarse sands and poor peat soils [82]. Extensive areas of poorly drained soils have also been linked with the absence of Japanese honeysuckle in southern Florida [15].

Anthropogenic effects on Japanese honeysuckle distribution are multifaceted. However, while roadways may facilitate the spread of the invader, artificial forest regeneration, one of the management practices we considered in this work, was shown effective in reducing the likelihood of Japanese honeysuckle invasions. A positive outcome of artificial regeneration, which generally involves some effort to control woody competition prior to planting, is that these types of sites are at least temporarily less susceptible to Japanese honeysuckle invasion. Approaches like herbicide application, mechanical treatment, or burning, aim to reduce interspecific competition and allow for successful establishment of the newly planted seedlings. However, as a side effect, this step may also reduce biodiversity on artificially regenerated sites in general (mean $H_s = 0.69$ *vs.* $H_s = 1.06$ on sites with and without history of artificial regeneration, respectively). The majority of the industrial plantations are located in southeastern Texas and, consequently, artificial regeneration is more common in this part of the state, with about 30% of the sites being classified in this category, as compared to about 21% of the sites in northeastern Texas, a region with less frequent human intervention and more favorable soil types.

We estimated invasion likelihood by applying boosted regression trees to data on landscape features, climatic conditions, forest conditions, disturbance factors, and forest management activities collected over a five-year period. Two valid criticisms of our approach are that our estimates of invasion likelihood are unique to our method of analysis [83] and to our specification of the variables included in that analysis [84]. These criticisms are generic problems related to structural uncertainty in the mathematical representation of natural systems [85]. We identified most of our potential independent variables following a recent study focused on the invasion of Japanese honeysuckle in forestlands of the southern United States [27]; however, we did not use exactly the same set of independent variables and we used a different modeling approach. The possibility remains that there might be a more powerful model [83] and/or a more useful set of explanatory variables [66]. Evaluation of the relative merits of the different methodological approaches to geographical distribution modeling currently is a topic of much debate [83,86,87], but is beyond the scope of the present study.

In conclusion, our analyses suggest that the range of Japanese honeysuckle in forestlands of eastern Texas is expanding. Even though providing reliable predictions of habitats that are most at risk, distribution limits, and efficacy of management strategies for any invasive species remain a challenge, our model showed the positive impact of selected management practices upon reducing the likelihood of invasion. Our model may also guide forest managers in development of long term monitoring

Forests **2015**, *6*, 4374–4390

and control strategies for Japanese honeysuckle by facilitating detection and eradication of newly established invasions in eastern Texas.

Acknowledgments: Acknowledgments: The study was supported by the Undergraduate Research Fund of the Department of Wildlife and Fisheries Sciences, Texas A&M University and The Pine Integrated Network: Education, Mitigation, and Adaptation Project (PINEMAP), a Coordinated Agricultural Project funded by the USDA National Institute of Food and Agriculture, Award #2011-68002-30185. We also thank three anonymous reviewers for their time and effort. The manuscript is greatly improved as a result of their comments.

Author Contributions: Author Contributions: This study was designed by Hsiao-Hsuan Wang. Erin K. McGrew organized the data. Tomasz E. Koralewski computed indexes. Hsiao-Hsuan Wang conducted statistical analyses. Hsiao-Hsuan Wang, Tomasz E. Koralewski, William E. Grant and Thomas D. Byram led the writing.

Conflicts of Interest: Conflicts of Interest: The authors declare no conflict of interest.

References

1. Moser, W.K.; Barnard, E.L.; Billings, R.F.; Crocker, S.J.; Dix, M.E.; Gray, A.N.; Ice, G.G.; Kim, M.-S.; Reid, R.; Rodman, S.U.; *et al.* Impacts of nonnative invasive species on US forests and recommendations for policy and management. *J. For.* **2009**, *107*, 320–327.

2. Pejchar, L.; Mooney, H.A. Invasive species, ecosystem services and human well-being. *Trends Ecol. Evol.* **2009**, *24*, 497–504. [CrossRef] [PubMed]

3. Pimentel, D.; Zuniga, R.; Morrison, D. Update on the environmental and economic costs associated with alien-invasive species in the United States. *Ecol. Econ.* **2005**, *52*, 273–288. [CrossRef]

4. Wilcove, D.S.; Rothstein, D.; Jason, D.; Phillips, A.; Losos, E. Quantifying threats to imperiled species in the United States. *BioScience* **1998**, *48*, 607–615. [CrossRef]

5. Gurevitch, J.; Padilla, D.K. Are invasive species a major cause of extinctions? *Trends Ecol. Evol.* **2004**, *19*, 470–474. [CrossRef] [PubMed]

6. Wang, H.-H.; Grant, W.E.; Gan, J.; Rogers, W.E.; Swannack, T.M.; Koralewski, T.E.; Miller, J.H.; Taylor, J.W. Integrating spread dynamics and economics of timber production to manage Chinese tallow invasions in southern US forestlands. *PLoS ONE* **2012**, *7*, e33877. [CrossRef] [PubMed]

7. McNulty, S.G.; Moore, J.A.; Iverson, L.; Prasad, A.; Abt, R.; Smith, B.; Sun, G.; Gavazzi, M.; Bartlett, J.; Murray, B.; *et al.* Application of linked regional scale growth, biogeography, and economic models for southeastern United States pine forests. *World Resour. Rev.* **2000**, *12*, 298–320.

8. Linder, E.T. *Biodiversity and Southern Forests*; General Technical Report SRS 75; US Department of Agriculture, Forest Service, Southern Research Station: Asheville, NC, USA, 2004; pp. 303–306.

9. Hayes, D. Earth day 1990: Threshold of the green decade. *World Policy J.* **1990**, *7*, 289–304.

10. Reubens, B.; Poesen, J.; Danjon, F.; Geudens, G.; Muys, B. The role of fine and coarse roots in shallow slope stability and soil erosion control with a focus on root system architecture: A review. *Trees* **2007**, *21*, 385–402. [CrossRef]

11. Schierenbeck, K.A. Japanese honeysuckle (*Lonicera japonica*) as an invasive species; history, ecology, and context. *Crit. Rev. Plant Sci.* **2004**, *23*, 391–400. [CrossRef]

12. Natural Resources Conservation Service (NRCS). The PLANTS Database. Available online: http://plants.usda.gov (accessed on 1 July 2015).

13. Hardt, R.A. Japanese honeysuckle: From "one of the best" to ruthless pest. *Arnoldia* **1986**, *46*, 27–34.

14. Andrews, E.F. The Japanese honeysuckle in the eastern United States. *Torreya* **1919**, *19*, 37–43.

15. Leatherman, A.D. Ecological Life-History of *Lonicera japonica* Thunb. Ph.D. Thesis, University of Tennessee, Knoxville, TN, USA, 1955.

16. Schierenbeck, K.A. Comparative ecological and genetic studies between a native (*Lonicera sempervirens* L.) and an introduced congener (*L. japonica* Thunb.). Ph.D. Thesis, Washington State University, Pullman, WA, USA, 1992.

17. West, N.M.; Gibson, D.J.; Minchin, P.R. Microhabitat analysis of the invasive exotic liana *Lonicera japonica* Thunb. *J. Torrey Bot. Soc.* **2010**, *137*, 380–390. [CrossRef]

18. Garrison, G.A.; Bjugstad, A.J.; Duncan, D.A.; Lewis, M.E.; Smith, D.R. *Vegetation and Environmental Features of Forest and Range Ecosystems*; Forest Service, Department of Agriculture: Washington, DC, USA, 1977; p. 68.

19. Nyboer, R. Vegetation management guideline—Japanese honeysuckle (*Lonicera japonica* Thunb.). *Nat. Areas J.* **1992**, *12*, 217–218.

20. Oosting, H.J.; Livingston, R.B. A resurvey of a loblolly pine community twenty-nine years after ground and crown fire. *Bull. Torrey Bot. Club* **1964**, *91*, 387–395. [CrossRef]

21. Yurkonis, K.A.; Meiners, S.J. Invasion impacts local species turnover in a successional system. *Ecol. Lett.* **2004**, *7*, 764–769. [CrossRef]

22. McLemore, B.F. *Comparison of Three Methods for Regenerating Honeysuckle-Infested Openings in Uneven-Aged Loblolly Pine Stands*; Southern Forest Experiment Station, Forest Service, US Department of Agriculture: Atlanta, GA, USA, 1985.

23. Wear, D.N.; Greis, J.G. Southern forest resource assessment: Summary of findings. *J. For.* **2002**, *100*, 6–14.

24. McKeand, S.; Mullin, T.; Byram, T.; White, T. Deployment of genetically improved loblolly and slash pines in the south. *J. For.* **2003**, *101*, 32–37.

25. United States Department of Agriculture (USDA). Southern Nonnative Invasive Plant data Extraction Tool (SNIPET). Available online: http://srsfia2.fs.fed.us/data_center/index.shtml (accessed on 1 July 2015).

26. Wang, H.-H.; Wonkka, C.L.; Grant, W.E.; Rogers, W.E. Potential range expansion of Japanese honeysuckle (*Lonicera japonica* Thunb.) in southern US forestlands. *Forests* **2012**, *3*, 573–590. [CrossRef]

27. Gan, J.; Smith, C.T. Co-benefits of utilizing logging residues for bioenergy production: The case for East Texas, USA. *Biomass Bioenergy* **2007**, *31*, 623–630. [CrossRef]

28. Rudis, V.A.; Gray, A.; McWilliams, W.; O'Brien, R.; Olson, C.; Oswalt, S.; Schulz, B. Regional Monitoring of Nonnative Plant Invasions with the Forest Inventory and Analysis program, Proceedings of the Sixth Annual FIA Symposium, Denver, CO, USA, 21–24 September 2006; McRoberts, R.E., Reams, G.A., Eds.; General Technical Report WO-70. USDA Forest Service: Denver, CO, USA, 2006; pp. 49–64.

29. Bechtold, W.A.; Patterson, P.L. *The Enhanced Forest Inventory and Analysis Program—National Sampling Design and Estimation Procedures*; General Technical Report SRS-80; Southern Research Station, Forest Service, US Department of Agriculture: Asheville, NC, USA, 2005; p. 14.

30. United States Department of Agriculture (USDA). *The Forest Inventory and Analysis Database: Database Description and Users Manual Version 5.1*; US Department of Agriculture Forest Service: Arlington, VA, USA, 2011; pp. 12–15.

31. Lemke, D.; Hulme, P.E.; Brown, J.A.; Tadesse, W. Distribution modelling of Japanese honeysuckle (*Lonicera japonica*) invasion in the Cumberland Plateau and Mountain Region, USA. *For. Ecol. Manag.* **2011**, *262*, 139–149. [CrossRef]

32. Lemke, D.; Schweitzer, C.J.; Tadesse, W.; Wang, Y.; Brown, J.A. Geospatial assessment of invasive plants on reclaimed mines in Alabama. *Invasive Plant Sci. Manag.* **2013**, *6*, 401–410. [CrossRef]

33. Daly, C.; Halbleib, M.; Smith, J.I.; Gibson, W.P.; Doggett, M.K.; Taylor, G.H.; Curtis, J.; Pasteris, P.P. Physiographically sensitive mapping of climatological temperature and precipitation across the conterminous United States. *Int. J. Climatol.* **2008**, *28*, 2031–2064. [CrossRef]

34. Belda, E.J.; Sánchez, A. Seabird mortality on longline fisheries in the western Mediterranean: Factors affecting bycatch and proposed mitigating measures. *Biol. Conserv.* **2001**, *98*, 357–363. [CrossRef]

35. Wills, C.; Condit, R.; Foster, R.B.; Hubbell, S.P. Strong density- and diversity-related effects help to maintain tree species diversity in a neotropical forest. *Proc. Natl. Acad. Sci. USA* **1997**, *94*, 1252–1257. [CrossRef] [PubMed]

36. Filipescu, C.N.; Comeau, P.G. Competitive interactions between aspen and white spruce vary with stand age in boreal mixedwoods. *For. Ecol. Manag.* **2007**, *247*, 175–184. [CrossRef]

37. Elith, J.; Leathwick, J.R.; Hastie, T. A working guide to boosted regression trees. *J. Anim. Ecol.* **2008**, *77*, 802–813. [CrossRef] [PubMed]

38. Ridgeway, G. Generalized Boosted Regression Models. Documentation on the R Package 'gbm', Version 1-5-7. Available online: http://www.i-pensieri.com/gregr/gbm.shtml (accessed on 4 April 2012).

39. Friedman, J.H. Stochastic gradient boosting. *Comput. Stat. Data Anal.* **2002**, *38*, 367–378. [CrossRef]

40. Hosmer, D.W.; Lemeshow, S. *Applied Logistic Regression*; John Wiley and Sons, Inc.: New York, NY, USA, 2000.

41. Morales, J.M.; Carlo, T.A. The effects of plant distribution and frugivore density on the scale and shape of dispersal kernels. *Ecology* **2006**, *87*, 1489–1496. [CrossRef]

42. Clark, J.S.; Lewis, M.; McLachlan, J.S.; HilleRisLambers, J. Estimating population spread: What can we forecast and how well? *Ecology* **2003**, *84*, 1979–1988. [CrossRef]

43. Davies, K.F.; Harrison, S.; Safford, H.D.; Viers, J.H. Productivity alters the scale dependence of the diversity-invasibility relationship. *Ecology* **2007**, *88*, 1940–1947. [CrossRef] [PubMed]
44. Baars, R.; Kelly, D. Survival and growth responses of native and introduced vines in New Zealand to light availability. *N. Z. J. Bot.* **1996**, *34*, 389–400. [CrossRef]
45. Carter, G.A.; Teramura, A.H. Vine photosynthesis and relationships to climbing mechanics in a forest understory. *Am. J. Bot.* **1988**, *75*, 1011–1018. [CrossRef]
46. Halls, L.K. *Flowering and Fruiting of Southern Browse Species*; US Department of Agriculture, Forest Service, Southern Forest Experiment Station: New Orleans, LA, USA, 1973; p. 10.
47. Blair, R.M. Growth and nonstructural carbohydrate content of southern browse species as influenced by light intensity. *J. Range Manag.* **1982**, *35*, 756–760. [CrossRef]
48. Elton, C.S. *The Ecology of Invasions by Plants and Anmials*; University of Chicago Press: Chicago, IL, USA, 1958.
49. Hector, A.; Dobson, K.; Minns, A.; Bazeley-White, E.; Hartley Lawton, J. Community diversity and invasion resistance: An experimental test in a grassland ecosystem and a review of comparable studies. *Ecol. Res.* **2001**, *16*, 819–831. [CrossRef]
50. Levine, J.M. Species diversity and biological invasions: Relating local process to community pattern. *Science* **2000**, *288*, 852–854. [CrossRef] [PubMed]
51. Lyons, K.G.; Schwartz, M.W. Rare species loss alters ecosystem function—Invasion resistance. *Ecol. Lett.* **2001**, *4*, 358–365. [CrossRef]
52. Levine, J.M.; D'Antonio, C.M. Elton revisited: A review of evidence linking diversity and invasibility. *Oikos* **1999**, *87*, 15–26. [CrossRef]
53. Wang, H.-H.; Grant, W.E. Determinants of Chinese and European privet (*Ligustrum sinense* and *Ligustrum vulgare*) invasion and likelihood of further invasion in southern US forestlands. *Invasive Plant Sci. Manag.* **2012**, *5*, 454–463. [CrossRef]
54. Richardson, D.M.; Pyšek, P.; Rejmánek, M.; Barbour, M.G.; Panetta, F.D.; West, C.J. Naturalization and invasion of alien plants: Concepts and definitions. *Divers. Distrib.* **2000**, *6*, 93–107. [CrossRef]
55. Mitchell, C.E.; Agrawal, A.A.; Bever, J.D.; Gilbert, G.S.; Hufbauer, R.A.; Klironomos, J.N.; Maron, J.L.; Morris, W.F.; Parker, I.M.; Power, A.G.; *et al*. Biotic interactions and plant invasions. *Ecol. Lett.* **2006**, *9*, 726–740. [CrossRef] [PubMed]
56. Pelczar, R. When vigor is not a virtue: Beware the beautiful but invasive Japanese honeysuckle. *Horticulture* **1995**, *73*, 64–66.
57. Sasek, T.W.; Strain, B.R. Implications of atmospheric CO_2 enrichment and climatic change for the geographical distribution of two introduced vines in the USA. *Clim. Chang.* **1990**, *16*, 31–51. [CrossRef]
58. Holle, B.V.; Simberloff, D. Ecological resistance to biological invasion overwhelmed by propagule pressure. *Ecology* **2005**, *86*, 3212–3218. [CrossRef]
59. Merritt, D.M.; Nilsson, C.; Jansson, R. Consequences of propagule dispersal and river fragmentation for riparian plant community diversity and turnover. *Ecol. Monogr.* **2010**, *80*, 609–626. [CrossRef]
60. Primack, R.B.; Miao, S.L. Dispersal can limit local plant distribution. *Conserv. Biol.* **1992**, *6*, 513–519. [CrossRef]
61. Tilman, D. Community invasibility, recruitment limitation, and grassland biodiversity. *Ecology* **1997**, *78*, 81–92. [CrossRef]
62. Moody, M.E.; Mack, R.N. Controlling the spread of plant invasions: The importance of nascent foci. *J. Appl. Ecol.* **1988**, *25*, 1009–1021. [CrossRef]
63. Sheley, R.L.; Mangold, J.M.; Anderson, J.L. Potential for successional theory to guide restoration of invasive plant-dominated rangeland. *Ecol. Monogr.* **2006**, *76*, 365–379. [CrossRef]
64. Miller, J.H.; Schelhas, J. Adaptive collaborative restoration: A key concept for invasive plant management. In *Invasive Plants and Forest Ecosystems*; Kohli, R.K., Singh, J.P., Eds.; CRC Press: Boca Raton, FL, USA, 2008; pp. 251–265.
65. Lodge, D.M.; Williams, S.; MacIsaac, H.J.; Hayes, K.R.; Leung, B.; Reichard, S.; Mack, R.N.; Moyle, P.B.; Smith, M.; Andow, D.A.; *et al.* Biological invasions: Recommendations for US policy and management. *Ecol. Appl.* **2006**, *16*, 2035–2054. [CrossRef]
66. Wang, H.-H.; Swannack, T.M.; Grant, W.E.; Gan, J.; Rogers, W.E.; Koralewski, T.E.; Miller, J.H.; Taylor, J.W. Predicted range expansion of Chinese tallow tree (*Triadica sebifera*) in forestlands of the southern United States. *Divers. Distrib.* **2011**, *17*, 552–565. [CrossRef]

67. Gelbard, J.L.; Belnap, J. Roads as conduits for exotic plant invasions in a semiarid landscape. *Conserv. Biol.* **2003**, *17*, 420–432. [CrossRef]

68. Parendes, L.A.; Jones, J.A. Role of light availability and sispersal in exotic plant invasion along roads and streams in the H.J. Andrews Experimental Forest, Oregon. *Conserv. Biol.* **2000**, *14*, 64–75. [CrossRef]

69. Schweitzer, J.A.; Larson, K.C. Greater morphological plasticity of exotic honeysuckle species may make them better invaders than native species. *J. Torrey Bot. Soc.* **1999**, *126*, 15–23. [CrossRef]

70. Pitts, T.D. Foods of eastern bluebirds during exceptionally cold weather in Tennessee. *J. Wildl. Manag.* **1979**, *43*, 752–754. [CrossRef]

71. Suthers, H.B.; Bickal, J.M.; Rodewald, P.G. Use of successional habitat and fruit resources by songbirds during autumn migration in central New Jersey. *Wilson Bull.* **2000**, *112*, 249–260. [CrossRef]

72. Halls, L.K. Managing deer habitat in loblolly-shortleaf pine forest. *J. For.* **1973**, *71*, 752–757.

73. Nixon, C.M.; McClain, M.W.; Russell, K.R. Deer food habits and range characteristics in Ohio. *J. Wildl. Manag.* **1970**, *34*, 870–886. [CrossRef]

74. Sotala, D.J.; Kirkpatrick, C.M. Foods of white-tailed deer, *Odocoileus virginianus*, in Martin County, Indiana. *Am. Midl. Nat.* **1973**, *89*, 281–286. [CrossRef]

75. Sheldon, J.J.; Causey, K. Deer habitat management—Availability use of Japanese honeysuckle by white-tailed deer. *J. For.* **1974**, *72*, 286–287.

76. Michael, E.D. Movements of white-tailed deer on the Welder Wildlife Refuge. *J. Wildl. Manag.* **1965**, *29*, 44–52. [CrossRef]

77. Inglis, J.M.; Hood, R.E.; Brown, B.A.; DeYoung, C.A. Home range of white-tailed deer in Texas coastal prairie brushland. *J. Mammal.* **1979**, *60*, 377–389. [CrossRef]

78. Thomas, J.W.; Teer, J.G.; Walker, E.A. Mobility and home range of white-tailed deer on the Edwards Plateau in Texas. *J. Wildl. Manag.* **1964**, *28*, 463–472. [CrossRef]

79. Sage, R.W., Jr.; Tierson, W.C.; Mattfeld, G.F.; Behrend, D.F. White-tailed deer visibility and behavior along forest roads. *J. Wildl. Manag.* **1983**, *47*, 940–953. [CrossRef]

80. Kilgo, J.C.; Labisky, R.F.; Fritzen, D.E. Influences of hunting on the behavior of white-tailed deer: Implications for conservation of the Florida panther. *Conserv. Biol.* **1998**, *12*, 1359–1364. [CrossRef]

81. Conner, W.; Stanturf, J.A.; Gardiner, E.S.; Schweitzer, C.J.; Ezell, A.W. Recognizing and overcoming difficult site conditions for afforestation of bottomland hardwoods. *Ecol. Restor.* **2004**, *22*, 183–193.

82. Jackson, L.W. Honeysuckles. *Shrubs and Vines for Northeastern Wildlife*; Gill, J.D., Healy, W.M., Eds.; US Department of Agriculture, Forest Service, Northeastern Forest Experiment Station: Upper Darby, PA, USA, 1974; pp. 71–82.

83. Elith, J.; Graham, C.H. Do they? How do they? WHY do they differ? On finding reasons for differing performances of species distribution models. *Ecography* **2009**, *32*, 66–77. [CrossRef]

84. Agresti, A. *An Introduction to Categorical Data Analysis*; John Wiley and Sons, Inc.: Hoboken, NJ, USA, 2007.

85. Walters, C.J. *Adaptive Management of Renewable Resources*; Macmillan Publishing Company: New York, NY, USA, 1986.

86. Barry, S.; Elith, J. Error and uncertainty in habitat models. *J. Appl. Ecol.* **2006**, *43*, 413–423. [CrossRef]

87. Elith, J.; Leathwick, J.R. Species distribution models: Ecological explanation and prediction across space and time. *Annu. Rev. Ecol. Evol. Syst.* **2009**, *40*, 677–697. [CrossRef]

forests

MDPI

Article

Benthic Collector and Grazer Communities Are Threatened by Hemlock Woolly Adelgid-Induced Eastern Hemlock Loss

Joshua K. Adkins [†] and Lynne K. Rieske *

Department of Entomology, University of Kentucky, S-225 Ag North, Lexington, KY 40546-0091, USA
* Author to whom correspondence should be addressed; lrieske@uky.edu;
  Tel.: +1-859-257-1167; Fax: +1-859-323-1120.
† Current address: Department of Biology, Transylvania University, 301 Brown Science Center,
  Lexington, KY 40508, USA; jadkins@transy.edu.

Academic Editor: Diana F. Tomback
Received: 5 May 2015; Accepted: 20 July 2015; Published: 6 August 2015

Abstract: Eastern hemlock (*Tsuga canadensis* (L.) is a foundation species in eastern North America where it is under threat from the highly invasive, exotic hemlock woolly adelgid (*Adelges tsugae*). Eastern hemlock is especially important in riparian areas of Central and Southern Appalachia, so we compared the spatial and temporal composition of benthic collector-gatherers, collector-filterers, and grazers in headwater streams with hemlock-dominated riparian vegetation to those with deciduous tree-dominated riparian vegetation to evaluate the extent to which adelgid-induced hemlock loss could influence composition and abundance of these two functional feeding groups. We found differences in benthic invertebrate abundance and family-level diversity based on riparian vegetation and sampling approach, and, often, riparian vegetation significantly interacted with location or season. Collector-gatherers and grazers were more abundant in eastern hemlock streams in the summer, when hemlock litter is readily available and deciduous litter is relatively sparse. Riparian eastern hemlock appears to exert considerable influence on benthic invertebrate functional feeding group composition in headwater stream communities, as expected with a foundation species. With the loss of eastern hemlock due to adelgid-induced mortality, we should expect to see alterations in spatial and temporal patterns of benthic invertebrate abundance and diversity, with potential consequences to both benthic and terrestrial ecosystem function.

Keywords: *Adelges tsugae*; *Tsuga canadensis*; riparian vegetation; headwater streams; benthic invertebrates; foundation species; invasive species

---

1. Introduction

Eastern hemlock (*Tsuga canadensis* (L.) Carrière; Pinales: Pinaceae) is a coniferous foundation species of eastern North American forests, and is a prominent component of riparian vegetation in Central Appalachia [1–3]. Foundation species, such as eastern hemlock, define much of the structure and function within a community by exerting a disproportionate influence on their surrounding environment, generating locally stable conditions and stabilizing fundamental ecosystem processes [4,5]. Eastern hemlock regulates nutrient cycling and stream base flows due to its persistent and elevated transpiration rates, and air, soil, and water temperatures beneath its dense canopy [6–9]. This canopy density significantly reduces light penetration, resulting in a paucity of understory associates [10]. Eastern hemlock needles and coarse woody debris decompose slowly, resulting in low rates of nitrogen mineralization and nitrification [11–13]. The slowly decomposing coarse woody debris generated from eastern hemlock remains prevalent in streams longer than that of deciduous tree and shrub species, creating in-stream microhabitats and altering sedimentation rates, flow dynamics, and nutrient cycling [12,14]. Fish community diversity in streams that drain eastern hemlock riparian zones is greater than streams with hardwood riparian zones [3]. Headwater streams with healthy eastern hemlock-dominated riparian zones support unique benthic invertebrate communities in New England [1,15], and, in Central Appalachia, the shredder component of these invertebrate communities is influenced by the presence of eastern hemlock [16]. Clearly, eastern hemlock exerts a measureable effect on an array of community and ecosystem interactions [5,17,18]. These communities are under threat by an invasive pest that is reshaping the riparian forests associated with these streams, the hemlock woolly adelgid (*Adelges tsugae* Annand; Hemiptera: Adelgidae).

Hemlock woolly adelgid is decimating eastern North American hemlocks [19–22], threatening headwater streams and potentially leading to extensive and permanent changes in the aquatic biota of these watersheds. Eastern hemlock exhibits little resistance to adelgid feeding and is not expected to regenerate, resulting in vegetative shifts to deciduous species [21,23]. Eastern hemlock mortality and the transition to hardwood dominance in headwater stream riparian areas could result in changes to the soil moisture regime with subsequent alterations to stream discharge, which may impact nutrient and carbon cycling in the southern Appalachians [8,9,22]. Benthic invertebrate community composition is correlated with surrounding forest composition [15,16,23], and streams draining eastern hemlock forests contain greater taxonomic richness than those associated with mixed deciduous forests [1]. Thus, changes in stream invertebrate communities due to adelgid-induced hemlock mortality could affect trophic interactions and energy cycling throughout the ecosystem.

Headwater streams supply energy downstream in the form of dissolved or particulate organic matter, nutrients, and benthic invertebrate prey items [24–27]. These small streams also export considerable energy to riparian areas in the form of emergent adult aquatic insects [28], which results in reciprocating energy dynamics between headwater streams and the riparian zone [29,30]. In temperate forests, headwater streams are often sustained by allochthonous coarse particulate organic matter (CPOM), which is converted to fine particulate organic matter (FPOM) via several pathways, one of which is by feeding activity of benthic shredders [31–33]. The resulting FPOM then serves as an energy resource for benthic collectors, which are further subdivided by their feeding mode; those that collect FPOM from the streambed are designated "collector-gatherers", whereas those that collect suspended FPOM from the water column are designated "collector-filterers" [34].

Benthic grazers, in contrast, utilize algae as a food resource, and since headwater streams are often densely shaded, primary productivity and grazer prevalence is often low in these habitats [31]. But since algal abundance and grazer abundance are intimately linked [35,36], any changes to riparian vegetation that alter stream light regimes and/or nutrient cycling could alter the complement of grazers in stream communities [35].

Collectors and grazers together represent half of the expected benthic invertebrate community within a hypothetical headwater stream [30]. Thus these two functional feeding groups represent a substantial pool of potential prey within aquatic and terrestrial food webs, which could be affected by hemlock woolly adelgid-induced loss of eastern hemlock [37].

We sought to gain a broad understanding of the extent to which adelgid-induced hemlock mortality might affect headwater stream communities, and focus here on benthic invertebrate collectors and grazers. We compare the spatial and temporal composition of collectors (both collector-gatherers and collector-filterers) and grazers over a two year period in headwater streams with hemlock dominated riparian vegetation to those with deciduous tree-dominated riparian vegetation, which represents one potential end point of the successional trajectory following adelgid-induced hemlock mortality [21]. This approach allows us to evaluate the extent to which adelgid-induced hemlock loss may influence the composition and abundance of these two functional feeding groups. Because of marked differences in stream inputs of allochthonous materials based on dominant riparian vegetation, we expected to find different communities of collectors in each stream type. This expectation is corroborated by previous findings that benthic shredders, which produce the FPOM utilized by collectors, are more abundant in eastern hemlock streams during summer months [16]. Finally, we expected that grazer abundance would be highest in deciduous streams prior to bud break and canopy formation, when shading would impede algal growth.

2. Results

2.1. Study Sites

Our sites (Figure 1) did not differ with respect to watershed area and were similar in stream depth, discharge, and elevation at the confluence between the two riparian zone vegetation designations [16]. Co-dominant canopy species based on basal area consisted of eastern hemlock, American beech (*Fagus grandifolia* Ehrhart), white oak (*Quercus alba* L.), tulip poplar (*Liriodendron tulipifera* L.), black birch (*Betula lenta* L.), red maple (*Acer rubrum* L.), and sourwood (*Oxydendrum arboretum* L), with a substantial *Rhododendron maximum* (L.) component in the understory. Riparian vegetation classified as deciduous-dominated contained significantly lower eastern hemlock basal area in the overstory (3.1 + 1.3 (S.E.) *versus* 12.6 + 2.2 (S.E.) m²/ha), but not the understory, relative to those classified as hemlock-dominated, but eastern hemlock basal area was similar across the three study locations [16].

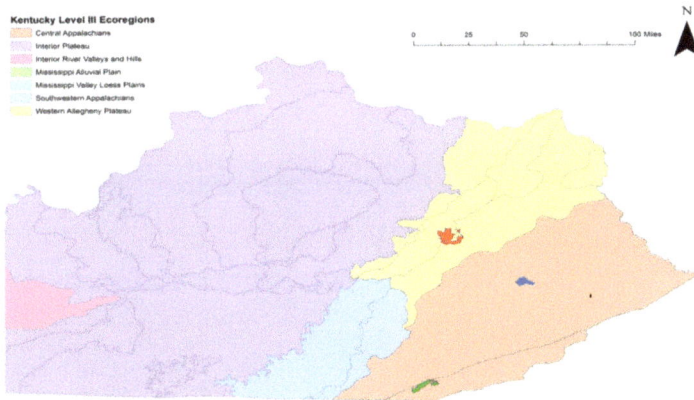

Figure 1. Location of sites in the Cumberland Plateau physiographic province of eastern Kentucky (USA) used to evaluate effects of dominant riparian vegetation on benthic collectors and grazers, including Red River Gorge (■), Robinson Forest (■), and Kentucky Ridge State Forest (■).

2.2. Benthic Invertebrate Collectors and Grazers

We collected 9532 invertebrates. Our sampling generated 28 collector taxa; 6 were classified as collector-filterers and 22 were classified as collector-gatherers. There were 9 taxa classified as grazers (Table 1).

The Hydropsychidae were most abundant collector-filterers in kick net samples ($n = 1347$), Surber samples ($n = 518$), and the Hester-Dendy samples ($n = 69$). The Chironomidae were the most common collector-gatherer (kick net and Hester-Dendy samples); in the Surber samples the Ephemerellidae were predominant ($n = 417$). The Heptageniidae were the most abundant grazers in kick net ($n = 1202$) and Surber samples ($n = 518$), but they were absent from Hester-Dendy samples. The most abundant grazer taxon in the Hester-Dendy samples was the Elmidae but only 4 individuals were collected (Table 1).

We found differences in benthic invertebrate abundance and family-level diversity based on riparian vegetation and sampling approach, and, often, riparian vegetation was involved in significant two-way interactions with location or season (Table S1). In particular, in samples collected via kick net we found significant ($p < 0.1$) vegetation by season interactions in the Elmidae, Shannon diversity index at the family level, and overall grazer abundance. In each instance values were elevated in hemlock streams during summer (Table 2 and Figure 2a–c). Additionally there was a weakly significant vegetation × season interaction for collector-gatherers sampled via kick nets, with the lowest abundance in summer. In Hester-Dendy samples, significant vegetation × season interactions were detected for the Heptageniidae, an Ephemeroptera categorized as a grazer, with abundance in summer greater in eastern hemlock streams than in their deciduous counterparts (Table 2 and Figure 2d). Collector-gatherer abundance collected via Hester-Dendy's in the fall was greater in hemlock streams than in deciduous streams (Table 2 and Figure 2e).

Table 1. Number of individuals from predominant collector-filterer, collector-grazer, and grazer families collected from kick net, Surber, and Hester-Dendy samples from eastern hemlock and deciduous dominated headwater streams, 2008–2010.

Functional Group	Family	Kick Net		Surber		Hester-Dendy		Total
		Deciduous	Hemlock	Deciduous	Hemlock	Deciduous	Hemlock	
	Hydropsychidae	612	735	252	266	36	33	1934
	Simuliidae	65	139	74	68	14	23	383
	Polycentropodidae	74	107	63	53	9	24	330
Collector-Filterers	Philopotamidae	44	36	8	10	1	4	103
	Isonychiidae	1	1	0	0	1	0	3
	Leptoceridae	0	3	0	0	0	0	3
	Brachycentridae	2	1	0	0	0	0	3
Total		798	1022	397	397	64	84	2759
	Chironomidae	300	447	111	91	90	78	1117
	Ephemerellidae	241	174	324	93	49	68	949
	Leptophlebiidae	159	219	58	51	20	18	525
	Baetidae	124	102	32	11	19	11	299
	Ameletidae	62	49	51	59	7	2	230
	Siphlonuridae	68	58	31	19	18	11	205
Collector-Gatherers	Ephemeridae	34	21	14	7	1	0	77
	Psychomyiidae	3	49	0	0	6	11	69
	Dixidae	12	15	4	2	11	7	51
	Caenidae	6	2	1	0	0	0	9
	Limnephilidae	0	0	4	5	0	0	9
	Rhyacophilidae	0	0	1	2	0	0	3
	Hydrophilidae	0	1	0	0	0	1	2
Total		1054	1182	634	340	232	213	3655
	Heptageniidae	608	594	690	316	44	43	2295
	Elmidae	139	121	57	41	4	4	366
	Uenoidae	36	43	26	43	1	3	152
	Psephenidae	58	36	24	15	0	1	134
Grazers	Dryopidae	12	41	6	16	3	13	91
	Glossosomatidae	11	13	13	6	0	2	45
	Goeridae	7	8	8	7	0	0	30
	Odontoceridae	0	0	0	0	0	4	4
	Helicopsychidae	0	1	0	0	0	0	1
Total		871	857	824	444	52	70	3118

Table 2. Abundance (mean (S.E.)) of benthic collectors and grazers collected in (a) kick nets; (b) Surber; and (c) Hester-Dendy Samplers from headwater streams with eastern hemlock or deciduous dominated riparian vegetation across two years in Kentucky (USA). For location, vegetation, and season, means within rows followed by the same letter are not significantly different ($\alpha = 0.1$).

	Vegetation		Location			Season		
	Deciduous	Hemlock	Kentucky Ridge	Robinson Forest	Red River Gorge	Fall	Spring	Summer
a. Kick net								
Chironomidae	1.0 (0.2) b	1.5 (0.3) a	0.4 (0.1) b	1.8 (0.4) a	1.6 (0.3) a	1.1 (0.4) b	2.3 (0.4) a	0.4 (0.1) b
Ephemerellidae	0.8 (0.2) a	0.6 (0.1) a	0.2 (0.1) a	0.6 (0.1) a	1.3 (0.3) a	0.3 (0.1) b	1.6 (0.3) a	0.1 (0.0) b
Leptophlebiidae	0.6 (0.1) b	0.8 (0.1) a	0.4 (0.1) a	0.6 (0.1) a	0.9 (0.1) a	1.0 (0.2) a	0.8 (0.1) a	0.2 (0.0) b
Siphlonuridae	0.2 (0.1) a	0.2 (0.1) a	0.3 (0.1) a	0.3 (0.1) a	0.1 (0.0) a	0.0 (0.0) b	0.5 (0.1) a	0.0 (0.0) b
Ameletidae	0.2 (0.1) a	0.2 (0.1) a	0.2 (0.1) a	0.4 (0.1) a	0.1 (0.0) a	0.0 (0.0) a	0.5 (0.1) a	0.0 (0.0) a
Ephemeridae	0.1 (0.0) a	0.1 (0.0) a	0.0 (0.0) a	0.0 (0.0) b	0.2 (0.1) a	0.2 (0.1) a	0.1 (0.0) a	0.0 (0.0) a
Psychomyiidae	0.0 (0.0) a	0.2 (0.1) a	0.0 (0.0) a	0.0 (0.0) a	0.2 (0.1) a	0.2 (0.2) a	0.1 (0.0) a	0.0 (0.0) a
Heptageniidae	2.1 (0.4) a	2.0 (0.3) b	1.4 (0.2) b	0.8 (0.2) b	3.7 (0.6) a	1.0 (0.1) b	4.4 (0.6) a	0.4 (0.1) b
Elmidae	0.5 (0.1) a	0.4 (0.1) b	0.1 (0.0) b	0.1 (0.0) b	1.0 (0.1) a	0.5 (0.1) ab	0.6 (0.1) a	0.3 (0.1) b
Psephenidae	0.2 (0.0) a	0.1 (0.0) a	0.2 (0.0) b	0.0 (0.0) c	0.3 (0.1) a	0.2 (0.0) a	0.2 (0.0) a	0.1 (0.0) a
Uenoidae	0.1 (0.0) a	0.1 (0.1) a	0.1 (0.1) a	0.0 (0.0) a	0.2 (0.1) a	0.0 (0.0) b	0.4 (0.1) a	0.0 (0.0) b
Dryopidae	0.0 (0.0) a	0.1 (0.0) a	0.0 (0.0) b	0.0 (0.0) b	0.2 (0.1) a	0.1 (0.0) a	0.0 (0.0) b	0.1 (0.1) a
Hydropsychidae	2.1 (0.3) a	2.5 (0.3) a	1.9 (0.3) b	0.3 (0.1) c	4.4 (0.5) a	3.4 (0.6) a	1.8 (0.3) b	2.0 (0.3) b
Simuliidae	0.2 (0.0) a	0.5 (0.1) a	0.2 (0.1) a	0.3 (0.1) a	0.5 (0.2) a	0.0 (0.0) b	0.9 (0.2) a	0.0 (0.0) b
Polycentropodidae	0.3 (0.1) a	0.4 (0.1) a	0.2 (0.0) b	0.1 (0.0) b	0.6 (0.1) a	0.3 (0.1) a	0.5 (0.1) a	0.1 (0.0) b
Philopotamidae	0.2 (0.1) a	0.1 (0.0) a	0.1 (0.0) a	0.2 (0.1) a	0.1 (0.1) a	0.2 (0.1) a	0.2 (0.1) a	0.0 (0.0) a
Simpson	0.5 (0.0) a	0.5 (0.0) a	0.5 (0.0) a	0.3 (0.0) b	0.6 (0.0) a	0.5 (0.0) a	0.6 (0.0) a	0.3 (0.0) b
Shannon	0.7 (0.0) b	0.8 (0.0) a	0.7 (0.0) b	0.5 (0.0) c	1.0 (0.0) a	0.8 (0.0) b	1.0 (0.0) a	0.5 (0.0) c
Evenness	0.6 (0.0) a	0.6 (0.0) a	0.6 (0.0) a	0.4 (0.0) b	0.7 (0.0) a	0.6 (0.0) a	0.7 (0.0) a	0.4 (0.0) b
Filterer	2.8 (0.3) a	3.5 (0.4) a	2.4 (0.3) b	0.8 (0.1) c	5.6 (0.6) a	3.9 (0.6) a	3.4 (0.4) a	2.2 (0.3) a
Gatherer	3.7 (0.5) a	4.1 (0.5) a	1.9 (0.2) b	4.2 (0.7) ab	5.2 (0.6) a	3.1 (0.5) b	7.1 (0.7) a	1.0 (0.1) c
Collector	6.4 (0.6) a	7.6 (0.8) a	4.4 (0.4) b	5.1 (0.8) b	10.8 (1.0) a	7.0 (1.0) b	10.4 (1.0) a	3.2 (0.4) b
Grazer	3.0 (0.4) a	2.9 (0.4) a	1.8 (0.2) b	1.0 (0.2) c	5.6 (0.7) a	1.9 (0.2) b	5.6 (0.7) a	1.0 (0.1) b
Total	9.5 (0.9) a	10.5 (1.0) a	6.2 (0.5) b	6.1 (0.9) b	16.5 (1.6) a	9.0 (1.1) b	16.0 (1.5) a	4.2 (0.5) b
b. Surber								
Ephemerellidae	2.3 (0.6) a	0.7 (0.1) a	0.2 (0.1) b	1.3 (0.3) ab	2.7 (0.7) a	0.1 (0.0) b	3.7 (0.7) a	0.1 (0.0) b
Chironomidae	0.8 (0.2) a	0.6 (0.1) a	0.1 (0.1) a	1.7 (0.3) a	0.3 (0.1) a	0.0 (0.0) b	1.7 (0.3) a	0.1 (0.0) b
Ameletidae	0.4 (0.1) a	0.4 (0.1) a	0.1 (0.1) a	0.9 (0.2) a	0.2 (0.1) a	0.0 (0.0) a	1.0 (0.2) a	0.0 (0.0) a
Leptophlebiidae	0.4 (0.1) a	0.4 (0.1) a	0.1 (0.0) b	0.4 (0.1) a	0.6 (0.2) a	0.5 (0.2) a	0.4 (0.1) a	0.3 (0.1) a
Siphlonuridae	0.2 (0.1) a	0.1 (0.1) a	0.1 (0.1) a	0.3 (0.1) a	0.1 (0.1) a	0.1 (0.0) a	0.4 (0.1) a	0.0 (0.0) a
Baetidae	0.2 (0.1) a	0.1 (0.0) a	0.0 (0.0) a	0.1 (0.0) a	0.3 (0.1) a	0.0 (0.0) b	0.3 (0.1) a	0.0 (0.0) b
Hydropsychidae	1.8 (0.3) a	1.9 (0.3) a	1.9 (0.4) b	0.2 (0.1) c	3.2 (0.4) a	2.3 (0.5) a	2.4 (0.4) a	0.9 (0.2) a
Simuliidae	0.5 (0.1) a	0.5 (0.3) a	0.2 (0.1) a	0.4 (0.1) a	0.8 (0.4) a	0.0 (0.0) b	1.2 (0.4) a	0.1 (0.0) b
Polycentropodidae	0.4 (0.1) a	0.4 (0.1) a	0.2 (0.1) b	0.1 (0.0) b	0.8 (0.2) a	0.4 (0.1) a	0.7 (0.2) a	0.0 (0.0) a
Heptageniidae	4.9 (1.3) a	2.2 (0.5) a	1.3 (0.2) a	1.5 (0.5) a	7.1 (1.7) a	0.7 (0.2) b	8.7 (1.7) a	0.2 (0.0) b
Elmidae	0.4 (0.1) a	0.3 (0.1) a	0.1 (0.0) a	0.3 (0.1) a	0.6 (0.1) a	0.2 (0.1) a	0.4 (0.1) a	0.4 (0.1) a
Simpson	0.4 (0.0) a	0.5 (0.0) a	0.4 (0.0) a	0.5 (0.0) b	0.6 (0.0) a	0.4 (0.0) b	0.7 (0.0) a	0.3 (0.0) b
Shannon	0.6 (0.0) a	0.7 (0.1) a	0.5 (0.1) b	0.7 (0.1) b	0.9 (0.1) a	0.6 (0.1) b	1.1 (0.1) a	0.4 (0.0) b
Evenness	0.5 (0.0) a	0.6 (0.0) a	0.5 (0.0) b	0.5 (0.0) b	0.6 (0.0) a	0.6 (0.0) a	0.7 (0.0) a	0.4 (0.0) b
Filterer	2.8 (0.4) a	2.8 (0.5) a	2.3 (0.5) b	0.8 (0.2) c	4.9 (0.7) a	2.8 (0.5) b	4.5 (0.7) a	1.1 (0.2) b
Gatherer	4.5 (0.8) a	2.4 (0.3) a	0.8 (0.2) b	4.8 (0.7) a	4.4 (0.9) a	1.0 (0.2) b	7.6 (1.0) a	0.8 (0.2) b
Collector	7.3 (1.1) a	5.2 (0.7) a	3.1 (0.5) b	5.7 (0.8) b	9.2 (1.4) a	3.8 (0.6) b	12.0 (1.4) a	1.9 (0.3) c
Grazer	5.8 (1.4) a	3.1 (0.5) b	1.5 (0.2) b	1.9 (0.5) b	8.9 (1.8) a	1.5 (0.2) b	9.9 (1.8) a	0.9 (0.2) b
Total	13.2 (2.3) a	8.4 (1.1) a	4.6 (0.7) b	7.6 (1.0) b	18.2 (3.1) a	5.3 (0.8) b	21.9 (3.0) a	2.8 (0.3) b
c. Hester-Dendy								
Chironomidae	0.5 (0.1) a	0.4 (0.1) a	0.2 (0.1) b	0.8 (0.2) a	0.5 (0.1) ab	0.2 (0.1) b	0.2 (0.2) a	0.2 (0.1) b
Ephemerellidae	0.3 (0.1) a	0.4 (0.1) a	0.2 (0.1) a	0.4 (0.1) a	0.4 (0.1) a	0.6 (0.2) a	0.1 (0.1) a	0.1 (0.0) a
Hydropsychidae	0.2 (0.0) a	0.2 (0.1) a	0.2 (0.1) ab	0.1 (0.0) b	0.3 (0.1) a	0.1 (0.1) b	0.1 (0.1) a	0.1 (0.0) b
Heptageniidae	0.2 (0.0) a	0.2 (0.0) b	0.2 (0.1) b	0.1 (0.0) c	0.4 (0.1) a	0.3 (0.1) a	0.1 (0.1) a	0.1 (0.0) b
Simpson	0.2 (0.0) a	0.2 (0.0) a	0.2 (0.0) a	0.1 (0.0) a	0.2 (0.0) a	0.2 (0.0) a	0.1 (0.0) a	0.1 (0.0) b
Shannon	0.2 (0.0) a	0.2 (0.0) a	0.2 (0.0) b	0.2 (0.0) b	0.3 (0.0) a	0.3 (0.0)	0.1 (0.0)	0.1 (0.0)
Evenness	0.2 (0.0) a	0.2 (0.0) a	0.2 (0.0) ab	0.2 (0.0) b	0.3 (0.0) a	0.3 (0.1) a	0.1 (0.0) a	0.1 (0.0) b
Filterer	0.3 (0.1) a	0.5 (0.1) a	0.2 (0.1) b	0.2 (0.1) b	0.8 (0.2) a	0.5 (0.2) a	0.1 (0.2) a	0.1 (0.0) b
Gatherer	1.3 (0.2) a	1.2 (0.2) a	0.7 (0.1) b	1.5 (0.3) a	1.5 (0.3) a	1.4 (0.2) a	0.6 (0.3) a	0.6 (0.1) b
Collector	1.6 (0.2) a	1.7 (0.3) a	0.9 (0.1) b	1.7 (0.3) ab	2.3 (0.4) a	1.8 (0.3) a	0.7 (0.4) a	0.7 (0.1) b
Grazer	0.3 (0.1) a	0.4 (0.1) a	0.3 (0.1) b	0.1 (0.0) c	0.6 (0.1) a	0.4 (0.1) a	0.3 (0.1) a	0.3 (0.1) a
Total	1.9 (0.3) a	2.0 (0.3) a	1.2 (0.2) b	1.9 (0.3) b	2.9 (0.4) a	2.3 (0.3) a	1.0 (0.4) a	1.0 (0.1) b

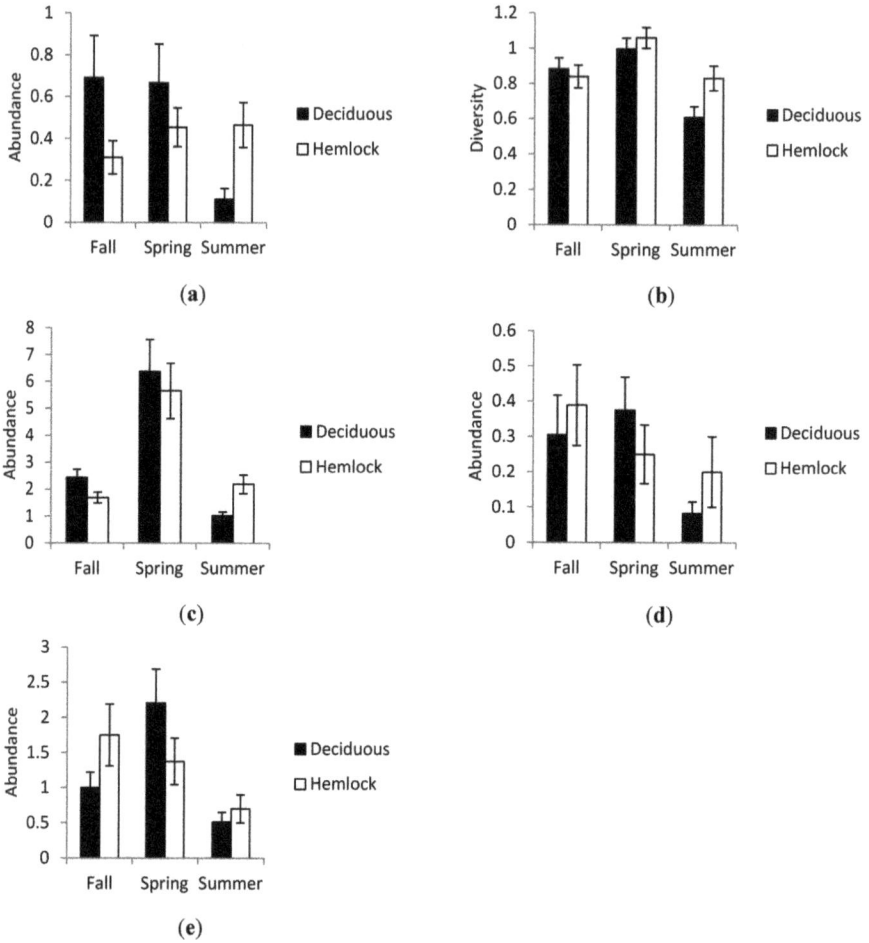

Figure 2. Interactions between riparian vegetation and season influences (**a**) abundance of Elmidae; (**b**) Shannon index of diversity; and (**c**) abundance of scrapers, sampled via kick nets, 2009–2010; and (**d**) Heptageniidae and (**e**) collector-gatherers, sampled via Hester-Dendy, fall 2009–summer 2010, from streams dominated by deciduous (■) and eastern hemlock (□) riparian vegetation.

The interacting effects of vegetation and location influenced the abundance of several taxa collected via kick net and Surbers, including the Chironomidae, Heptageniidae, Elmidae, Psephenidae, and Uenoidae (Table S1). This interaction was detected in grazers and total collectors (gatherers and filterers combined); in general these groups were more abundant in deciduous streams at Red River Gorge. The Chironomidae are an exception; the greatest abundance of chironomids was found at Robinson Forest in deciduous streams (Table 2).

2.3. Multivariate Ordination

The multivariate ordination indicated that headwater stream benthic invertebrate functional feeding group abundance was influenced by riparian vegetation and stream chemistry. The Monte Carlo test was significant ($p < 0.05$) for each ordination axis when all environmental variables (see

Table S2) were incorporated into the CCA (Table 3), indicating linear relationships between feeding groups and the environmental data for each ordination axis [38].

Table 3. Canonical correspondence analysis of benthic invertebrates collected via kick net and environmental parameters from streams dominated by eastern hemlock and deciduous riparian vegetation.

	Ordination Axis		
	1	2	3
Monte Carlo Test—Taxa-Environment Correlations	$p = 0.04$	$p = 0.02$	$p = 0.02$
Eigenvalues	0.018	0.014	0.006
Pearson species-environment correlations	0.392	0.318	0.305
Cumulative percentage of variance of species data	4.6	8.2	9.8
Intraset correlation for environmental variables			
Understory eastern hemlock stems	0.522	−0.122	−0.22
Understory deciduous stems	0.212	0.336	−0.233
Overstory eastern hemlock stems	0.005	−0.079	0.257
Overstory deciduous stems	0.058	−0.241	0.035
Sulfate	0.276	−0.812	−0.192
Nitrate	−0.513	−0.174	−0.334
Ammonia	−0.123	−0.067	0.351
Total Phosphorus	−0.023	−0.026	−0.292
Total Carbon	−0.214	0.113	0.142
Dissolved Organic Carbon	−0.068	0.197	0.121
pH	0.136	−0.072	0.222
Dissolved Oxygen	0.224	−0.086	0.141
Conductivity	0.147	−0.299	0.152
Functional feeding group weights			
Shredder	0.241	0.04	0.283
Scraper	−0.185	0.499	−0.258
Collector-Gatherer	0.484	−0.147	−0.092
Collector-Filterer	−0.272	0.194	0.648
Predator	−0.447	−0.475	−0.11

The CCA ordination accounted for 9.8% of the overall variation within the data (Table 3). The first ordination axis accounted for 4.6% of the variance; this axis was most strongly linked to understory hemlock stem count, and negatively associated with stream nitrate concentrations (Table 3, Figure 3). Collector-gatherers and shredders were positively associated with axis 1, while there was a negative association for scrapers, collector-filterers, and predators (Table 3). The second ordination axis accounted for 3.6% of the overall variation. The strongest positive association with this axis was understory deciduous stems, and sulfate concentrations generated the strongest negative association (Table 3, Figure 3a). Scrapers had the strongest positive association to this ordination axis, and predators, followed by collector-gatherers, had the strongest negative association (Table 3). Finally, 1.6% of the overall variation was accounted for by the third ordination axis. The environmental variable most strongly tied to this axis was stream ammonia concentrations (Table 3, Figure 3b). Positive associations with axis 3 were found for shredders and collector-filterers, while scrapers, collector-filterers, and predators were negatively associated (Table 3).

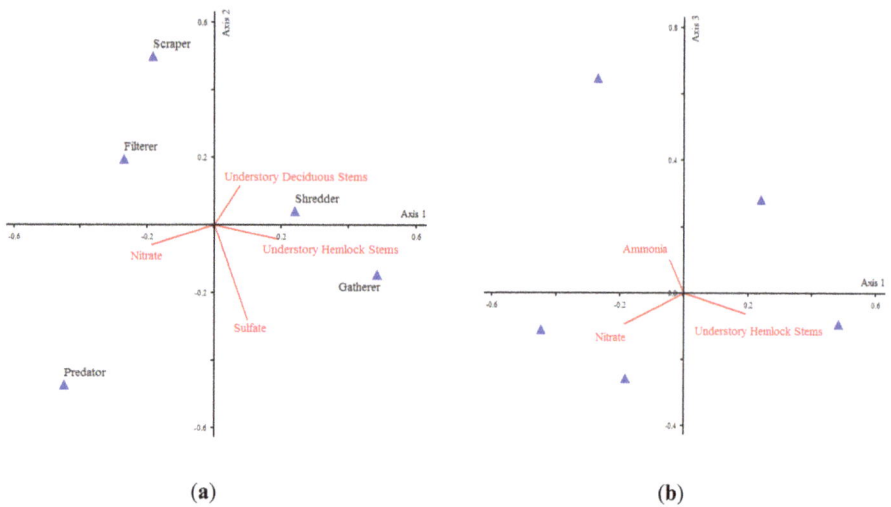

(a)   **(b)**

Figure 3. Canonical correspondence analysis showing (**a**) ordination axis 1 and 2; and (**b**) ordination axis 1 and 3 for benthic invertebrate functional feeding groups collected from headwater streams with riparian zones dominated by eastern hemlock and deciduous trees at the Robinson Forest site in 2010.

3. Discussion

We sought to understand the extent to which eastern hemlock riparian vegetation influences benthic invertebrate community composition in headwater streams. To this end we compared benthic collector and grazer community parameters in streams with eastern hemlock dominated riparian vegetation to those same parameters in streams with deciduous dominated riparian vegetation to evaluate the extent to which adelgid-induced hemlock loss could influence composition and abundance of these two functional feeding groups. This comparative approach has been used extensively to characterize the influence of eastern hemlock in structuring forest communities [1,3,15,16,20,38–40].

Eastern hemlock is prevalent in headwater riparian zones of central and southern Appalachia, and while there was some eastern hemlock present in our streams classified as deciduous dominated, eastern hemlock stem density was greater in streams that were categorized as eastern hemlock. We found distinct differences in associated benthic invertebrate communities throughout the year. We found only minor differences in stream chemistry between the two stream types ([41] and Table S2), corroborating studies that found similarities in temperature, discharge, pH, or nitrate between eastern hemlock and deciduous dominated streams [1,42]. The water chemistry parameters for our study streams approximate those reported in undisturbed Appalachian streams [43], and these streams are considered high quality via the Hilsenhoff Family Index of Biotic Integrity [42,44,45].

Riparian vegetation alone did not affect any measure of benthic collector or grazer evenness or family level diversity, in contrast to similar studies in the Northeastern US [1,15]. Snyder *et al.* [1] found that eastern hemlock headwater streams had greater invertebrate diversity (via Simpson's Index) than did streams with deciduous dominated riparian vegetation, whereas Willacker *et al.* [15] found greater richness, more unique taxa, and more overall taxa in deciduous dominated streams relative to hemlock dominated streams, but also found higher collector-gatherers in hemlock-dominated streams. Benthic invertebrate abundance and richness is likely linked to both litter quality and seasonal availability. We found that collector-gatherers and grazers were more abundant in eastern hemlock streams in the summer, a time of year when hemlock litter is readily available and deciduous tree litter is relatively sparse. Eastern hemlock litter enters streams more consistently, though in smaller amounts, than does litter from deciduous trees [16]. Conifer litter is considered low quality;

hemlock litter is low in nitrogen [11] with low rates of microbial conditioning [46]. Shredders also comprise a sizeable portion of the benthic invertebrate communities in these streams, and show seasonal fluctuations in abundance that reflects seasonal fluctuations in hemlock litter inputs [16]. The consistent input and slow rate of microbial conditioning of eastern hemlock results in a constant and available resource for shredders, which generate FPOM through their feeding activities, thereby providing abundant resources that facilitate growth and development of collector-gatherers during summer months [31,47–50]. Shredder-collector-grazer processing chains have been demonstrated experimentally [49,51–53], and suggest that changes in leaf-litter contributions to streams can alter the abundance of benthic functional feeding groups [24], with repercussions for stream and riparian food webs and energy flow [54–57].

Our multivariate analysis provides insights into environmental factors that contribute to the observed patterns in invertebrate functional feeding group communities. The CCA explains 9.8% of the overall variation within our data, typical of the <10% variation commonly accounted for in ecological data using this approach [58]. Our CCA corroborates the significant findings of the generalized linear mixed model for functional feeding groups, and demonstrates that shredders and collector-gatherers are positively associated with understory eastern hemlock stem density, which occurs at greater densities in streams classified as eastern hemlock dominated [16,41]. Riparian eastern hemlock appears to exert considerable influence on benthic invertebrate functional feeding group composition in headwater stream communities, as expected with a foundation species. With the loss of eastern hemlock due to adelgid-induced mortality, we should expect to see alterations in spatial and temporal patterns of benthic invertebrate abundance and diversity.

4. Materials and Methods

4.1. Study Sites

Three protected areas in the Cumberland Plateau physiographic province of eastern Kentucky were selected (Figure 1). This region contains steep, mountainous terrain with underlying shale and sandstone and abundant coal seams [59]. The Red River Gorge Geological Area and Natural Bridge State Park State Nature Preserve is located in the Northern Forested Plateau Escarpment ecoregion. Robinson Forest is situated in the Dissected Appalachian Plateau ecoregion, and Kentucky Ridge State Forest is located further south in the Cumberland Mountain Thrust Block ecoregion [60]. Yearly precipitation ranges from 106 to 139 cm, and temperatures can range from −6.2 to 8.3 °C in January and from 16.6 to 31.6 °C in July [60]. Elevation ranges from 167 to 1261 m throughout this area of the state [60]. The dominant vegetation type is mixed mesophytic forest [61,62] and eastern hemlock is found throughout [63].

Candidate streams were selected using GIS and remote sensing [42], and the presence of riparian eastern hemlock was determined using the vegetation database from the Kentucky GAP Analysis [64]. Boundaries of drainage basins were determined using surface hydrology modeling of 30 m resolution digital elevation models from the Kentucky Office of Geographic Information. Drainage basin areas were extracted from digital elevation models using the Hydrology toolkit of ArcGIS [65]. Streams with similar drainage areas and suitable eastern hemlock in the riparian zone were visited and evaluated for suitability. Ultimately, three streams with riparian zones dominated by eastern hemlock and three streams with deciduous dominated riparian zones were selected at each of the three sites for a total of eighteen streams. A 30 m reach was designated in each stream at least 150 m upstream of the confluence for vegetation and stream characterization and benthic invertebrate sampling [16].

We assessed vegetative composition and structure using two 0.04-ha fixed-radius whole plots randomly placed in accessible areas within each stream's riparian zone, one on each side of the stream [16,41]. Ten subplots, five 0.004-ha and five 0.0004-ha, were nested within each whole plot to enhance precision of the vegetation assessments. Whole plots were utilized to assess overstory and midstory vegetation, 0.004-ha subplots were used to assess saplings and shrubs (>137 cm height), and 0.0004-ha micro plots were used to assess seedlings, shrubs (<137 cm height), and vines. One of each subplot was positioned at the whole plot center and in each cardinal direction, 7.7 m from the plot center. Each surveyed reach contained two 0.04-ha whole plots, ten 0.004-ha subplots, and ten 0.0004-ha microplots [66]. Measurements of vegetation and plot data followed the Common Stand Exam protocol of the USDA Forest Service's Natural Resource Information System: Field Sampled Vegetation Module [67].

Watershed areas for the 18 streams were compared between dominant riparian vegetation type and across study locations using analysis of variance with a Tukey adjustment (PROC GLM, SAS Software v9.3). The number of stems and basal area of eastern hemlock and the six most commonly encountered overstory deciduous trees were calculated from riparian vegetation assessments and compared between eastern hemlock dominated and deciduous dominated riparian zones using the Kruskal-Wallis non-parametric procedure (PROC NPAR1WAY). High, median, and low temperature readings from each sample date were analyzed using analysis of variance between dominant riparian vegetation type and across locations. Comparison of least squares means was used as a post-hoc means separation procedure when appropriate. All analyses were performed using SAS v9.3 [68].

4.2. Benthic Invertebrate Collectors and Grazers

Three cross-stream transects were established across three riffles within each 30 m reach, and benthic invertebrates were sampled at 30 d intervals using a standard kick net (30 cm wide by 17 cm tall) with a 30-second kick interval and a 0.25 m² Surber sampler. Each of the three riffles per stream reach were sampled for a total six samples per stream (n = 54 for each sampling method). Artificial substrates (Hester-Dendy samplers, Forestry Suppliers; Jackson, MS, USA) consisting of five 2.5 × 5 cm² plates were used to passively monitor colonization by invertebrates; one sampler was deployed at each end of the designated 30 m reach (two per stream, n = 18 per riparian vegetation type). The use of multiple collection approaches allowed us to sample benthic invertebrates that utilize different microhabitats within the lotic environment, to provide a more complete picture of the benthic collector and grazer communities. Benthic invertebrate samples were preserved in the field using 70% ETOH and identified in the laboratory to the order or family level [69]. Each individual specimen was then assigned to a functional feeding group based on the most common feeding mode in each family (collector-gatherer, collector-filterer, grazer, shredder, or predator) [69]. Monthly sampling began in September 2008 and concluded in September 2010.

Three sampling intervals that represent spring, summer, and fall across two years were used to assess the influence of riparian eastern hemlock on benthic collectors and grazers from kick net and Hester-Dendy samples (see Supplemental Materials). Only one year's data are available for the Surber sampler, and only one set of fall data are available for the Hester-Dendy samples due to low stream flow. Shannon's and Simpson's diversity indices were calculated based on family level identifications [41]. Invertebrate abundances across the three sampling methods were calculated and compared by riparian vegetation type, study location, and season using a generalized linear mixed model with a split-plot design. Tukey's HSD (α = 0.1) was used as a *post hoc* means separation procedure when appropriate. Only those taxa with abundances greater than 50 were considered for the analysis.

4.3. Multivariate Ordination

The influence of riparian overstory and understory vegetation and stream chemistry characteristics on benthic macroinvertebrate abundance was evaluated using canonical correspondence analysis (CCA), which allowed us to explore arthropod associations along environmental gradients [70] to provide insight as to which stream characteristics most influence benthic collector and grazer populations. CCA is widely used by ecologists to explore and relate patterns of taxa distribution or abundance to environmental variables [38,58,70,71]. This approach provides an indication of which of the environmental variables included in an analysis are important in structuring community composition [38,58]. Only invertebrates from the upstream and downstream-most sample riffles

Forests **2015**, *6*, 2719–2738

were used for this analysis, as they corresponded with the upstream and downstream vegetation assessment plots; the midstream riffle was disregarded. In streams with riparian zones classified as either hemlock or deciduous, we compared invertebrate abundance to overstory and understory eastern hemlock and deciduous stem counts, and incorporated stream physical and chemical variables including concentrations of sulfate, nitrate, ammonia, total phosphorus, calcium and magnesium ions, and dissolved oxygen, pH, and conductivity [42]. A Monte Carlo permutation with 300 iterations was used to evaluate the influence of random events on the relationship between environmental variables and taxa abundance [72]. This procedure was performed solely on the kick net dataset as it was the most robust.

Significant Monte Carlo tests ($p < 0.1$) indicate the formation of linear combinations of environmental variables that maximally separate the niches of the taxa that are present [58] between benthic invertebrate functional feeding groups and riparian vegetation for the ordination axes [38]. We present the strongest intraset correlation values, which indicate the environmental variables with the greatest correlation to functional feeding group abundances [71]. Weights demonstrate the association of invertebrate taxa with the ordination axes, and eigenvalues explain variance extracted in relation to environmental variables. We used biplots to present relationships of functional feeding groups to riparian vegetation [38,71,73]. Environmental gradients include stem density of riparian hemlocks or deciduous trees and stream chemical parameters, and are characterized as lines radiating from the center of the plot, the length and direction of which relate to the strength of the relationships between environmental variables [70].

5. Conclusions

We compared the spatial and temporal composition of benthic collector-gatherers, collector-filterers, and grazers in headwater streams with hemlock dominated riparian vegetation to those with deciduous riparian vegetation to evaluate the extent to which adelgid-induced hemlock loss could influence composition and abundance of these two functional feeding groups. Consistent with expectations, we found elevated family level diversity and abundance of collector-gatherers in Hester-Dendy samples from hemlock streams collected in summer, potentially attributable to the steady input of hemlock litter providing food resources for benthic shredders [16], which then create FPOM for use by collector-gathers [31]. Contrary to expectations we also found increased abundance of grazing heptageniids in hemlock streams during summer sampling; we had expected grazer abundance to be higher in deciduous streams during spring, before canopy formation impeded algal growth. Riparian vegetation provides linkages between streams, influences headwater stream conditions, serves to support food webs, and affects community structure of downstream habitats [32,46,47,50,51]. Changes in headwater benthic invertebrate communities should be expected as hemlock woolly adelgid continues to invade the range of eastern hemlock in North America.

Acknowledgments: The authors thank Josh Adams, Paul Ayayee, Daniel Bowker, Josh Clark, Tom Coleman, Zachary Cornett, Murphey Coy, Joe Hacker, Cat Hoy, Amber Jones, Tom Kuhlman, Rachael Mallis, Abraham Nielsen, Heather Spaulding, Melanie Sprinkle, Matt Thomas, and Sarah Wightman for field and laboratory assistance. Luke Dodd and Xia Yu provided statistical guidance, and Chris Barton, John Obrycki and Lee Townsend reviewed an earlier version of this manuscript. We appreciate the comments of three anonymous reviewers, whose suggestions greatly improved this work. We also thank the University of Kentucky Robinson Forest Field Station, USDA Forest Service, Kentucky Division of Forestry, and the Kentucky State Nature Preserves Commission. This project was funded in part by grant funds from the Kentucky Water Resources Research Institute and the Tracy Farmer Center for the Environment and through funds provided by McIntire Stennis, and is published as Kentucky Agricultural Experiment Station publication number 15-08-052 with the approval of the Director.

Author Contributions: Co-authors shared equally in the conceptualization, interpretation, and presentation of this work. Data collection and data analysis was performed by J.K. Adkins.

Conflicts of Interest: The authors declare no conflict of interest.

References

1. Snyder, C.D.; Young, J.A.; Lemarié, D.P.; Smith, D.R. Influence of eastern hemlock (*Tsuga canadensis*) forests on aquatic invertebrate assemblages in headwater streams. *Can. J. Fish. Aquat. Sci.* **2002**, *59*, 262–275. [CrossRef]
2. Vandermast, D.B.; van Lear, D.H. Riparian vegetation in the southern Appalachian mountains (USA) following chestnut blight. *For. Ecol. Manag.* **2002**, *155*, 97–106. [CrossRef]
3. Ross, R.M.; Bennett, R.M.; Snyder, C.D.; Young, J.A.; Smith, D.R.; Lemarie, D.P. Influence of eastern hemlock (*Tsuga canadensis* L.) on fish community structure and function in headwater streams of the Delaware River basin. *Ecol. Freshw. Fish* **2003**, *12*, 60–65. [CrossRef]
4. Dayton, P.K. Toward an understanding of community resilience and the potential effects of enrichment to the benthos at McMurdo Sound, Antarctica. In Proceedings of the Colloquium on Conservation Problems in Antarctica, Blacksburg, VA, USA, 10–12 September 1971; Parker, B.C., Ed.; Allen Press: Lawrence, KS, USA, 1972; pp. 81–95.
5. Ellison, A.M.; Bank, M.S.; Clinton, B.D.; Colburn, E.A.; Elliott, K.; Ford, C.R.; Foster, D.R.; Kloeppel, B.D.; Knoepp, J.D.; Lovett, G.M.; *et al.* Loss of foundation species: Consequences for the structure and dynamics of forested ecosystems. *Front. Ecol. Environ.* **2005**, *3*, 479–486. [CrossRef]
6. Yorks, T.E.; Jenkins, J.C.; Leopold, D.J.; Raynal, D.J.; Orwig, D.A. Influences of eastern hemlock mortality on nutrient cycling. In Proceedings of the Symposium on Sustainable Management of Hemlock Ecosystems in Eastern North America, Durham, NH, USA, 22–24 June 1999; McManus, K., Shields, K., Souto, D., Eds.; Northeastern Research Station: Newtown Square, PA, USA, 2000; pp. 126–133.
7. Welsh, H.H.; Droege, S. A case for using plethodontid salamanders for monitoring biodiversity and ecosystem integrity of North American forests. *Conserv. Biol.* **2001**, *15*, 558–569. [CrossRef]
8. Ford, C.R.; Vose, J.M. *Tsuga canadensis* (L.) Carr. mortality will impact hydrologic processes in southern Appalachian forest ecosystems. *Ecol. Appl.* **2007**, *17*, 1156–1167. [CrossRef] [PubMed]
9. Nuckolls, A.E.; Wurzburger, N.; Ford, C.R.; Hendrick, R.L.; Vose, J.M.; Kloeppel, B.D. Hemlock declines rapidly with hemlock woolly adelgid infestation: Impacts on the carbon cycle of Southern Appalachian forests. *Ecosystems* **2008**, *12*, 179–190. [CrossRef]
10. Rankin, W.; Tramer, E.J. The gap dynamics of canopy trees of a *Tsuga canadensis* forest community. *Northeast. Nat.* **2002**, *9*, 391–406. [CrossRef]
11. Finzi, A.; Breemen, N.V. Canopy tree-soil interactions within temperate forests: Species effects on soil carbon and nitrogen. *Ecol. Appl.* **1998**, *8*, 440–446.
12. Jenkins, J.C.; Aber, J.D.; Canham, C.D. Hemlock woolly adelgid impacts on community structure and N cycling rates in eastern hemlock forests. *Can. J. For. Res.* **1999**, *29*, 630–645. [CrossRef]
13. Yorks, T.E.; Leopold, D.J.; Raynal, D.J. Effects of *Tsuga canadensis* mortality on soil water chemistry and understory vegetation: Possible consequences of an invasive insect herbivore. *Can. J. For. Res.* **2003**, *33*, 1525–1537. [CrossRef]
14. Pitt, D.B.; Batzer, D.P. Potential impacts on stream macroinvertebrates of an influx of woody debris from eastern hemlock demise. *For. Sci.* **2015**, *61*. [CrossRef]
15. Willacker, J.J.; Sobczak, W.V.; Colburn, E.A. Stream macroinvertebrate communities in paired hemlock and deciduous watersheds. *Northeast. Nat.* **2009**, *16*, 101–112. [CrossRef]
16. Adkins, J.K.; Rieske, L.K. A terrestrial invader threatens a benthic community: Hemlock woolly adelgid-induced loss of eastern hemlock alters invertebrate shredders in headwater streams. *Biol. Invasions* **2015**, *17*, 1163–1179. [CrossRef]
17. Webster, J.R.; Morkeski, K.; Wojculewski, C.A.; Niederlehner, B.V.; Benfield, E.F.; Elliott, K.J. Effects of hemlock mortality on streams in the southern Appalachian mountains. *Am. Midl. Nat.* **2012**, *168*, 112–131. [CrossRef]
18. Northington, R.M.; Webster, J.R.; Benfield, E.F.; Cheever, B.M.; Niederlehner, B.R. Ecosystem function in Appalachian headwater streams during an active invasion by the hemlock woolly adelgid. *PLoS ONE* **2013**. [CrossRef] [PubMed]
19. McClure, M.S. Density-dependent feedback and population cycles in *Adelges tsugae* (Homoptera: Adelgidae) on *Tsuga canadensis*. *Environ. Entomol.* **1991**, *20*, 258–264. [CrossRef]

20. Rohr, J.R.; Mahan, C.G.; Kim, K.C. Response of arthropod biodiversity to foundation species declines: The case of the eastern hemlock. *For. Ecol. Manag.* **2009**, *258*, 1503–1510. [CrossRef]

21. Spaulding, H.L.; Rieske, L.K. The aftermath of an invasion: Structure and composition of Central Appalachian hemlock forests following establishment of the hemlock woolly adelgid, *Adelges tsugae. Biol. Invasions* **2010**, *12*, 3135–3143. [CrossRef]

22. Ford, C.R.; Elliott, K.J.; Clinton, B.D.; Kloeppel, B.D.; Vose, J.M. Forest dynamics following eastern hemlock mortality in the southern Appalachians. *Oikos* **2012**, *121*, 523–536. [CrossRef]

23. Orwig, D.A.; Foster, D.R. Forest response to the introduced hemlock woolly adelgid in southern New England, USA. *J. Torrey Bot. Soc.* **1998**, *125*, 60–73. [CrossRef]

24. Wallace, A.J.B.; Eggert, S.L.; Meyer, J.L.; Webster, J.R.; Wallace, J.B. Multiple trophic levels of a forest stream linked to terrestrial litter inputs. *Science* **1997**, *277*, 102–104. [CrossRef]

25. Gregory, S.V.; Swanson, F.J.; McKee, W.A.; Cummins, K.W. An ecosystem perspective of riparian zones. *Bioscience* **1991**, *41*, 540–551. [CrossRef]

26. Naiman, R.J.; Décamps, H. The ecology of interfaces: Riparian zones. *Annu. Rev. Ecol. Syst.* **1997**, *28*, 621–658. [CrossRef]

27. Wipfli, M.S.; Gregovich, D.P. Export of invertebrates and detritus from fishless headwater streams in southeastern Alaska: Implications for downstream salmonid production. *Freshw. Biol.* **2002**, *47*, 957–969. [CrossRef]

28. Jackson, J.K.; Fisher, S.G. Secondary production, emergence, and export of aquatic insects of a Sonoran Desert stream. *Ecology* **1986**, *67*, 629–638. [CrossRef]

29. Fukui, D.; Murakami, M.; Nakano, S.; Aoi, T. Effect of emergent aquatic insects on bat foraging in a riparian forest. *J. Anim. Ecol.* **2006**, *75*, 1252–1258. [CrossRef] [PubMed]

30. Moldenke, A.R.; ver Linden, C. Effects of clearcutting and riparian buffers on the yield of adult aquatic macroinvertebrates from headwater streams. *For. Sci.* **2007**, *53*, 308–319.

31. Vannote, R.L.; Minshall, G.W.; Cummins, K.W.; Sedell, J.R.; Cushing, C.E. The river continuum concept. *Can. J. Fish. Aquat. Sci.* **1980**, *37*, 130–137. [CrossRef]

32. Knight, A.W.; Bottorff, R.L. The importance of riparian vegetation to stream ecosystems. In *California Riparian Systems, Ecology, Conservation, and Productive Management*; Warner, R.E., Hendrix, K.M., Eds.; University of California Press: Berkeley, CA, USA, 1984; pp. 160–167.

33. Gomi, T.; Sidle, R.C.; Richardson, J.S. Understanding processes and downstream linkages of headwater systems. *Bioscience* **2002**, *52*, 905–916. [CrossRef]

34. Cummins, K.W.; Wilzbach, M.A.; Gates, D.M.; Perry, J.B.; Taliaferro, W.B. Shredders and riparian vegetation. *Bioscience* **1989**, *39*, 24–30. [CrossRef]

35. Rowell, T.J.; Sobczak, W.V. Will stream periphyton respond to increases in light following forecasted regional hemlock mortality? *J. Freshw. Ecol.* **2008**, *23*, 33–40. [CrossRef]

36. Gregory, S.V. Plant-herbivore interactions in stream systems. In *Stream Ecology: Application and Testing of General Ecological Theory*; Barnes, J.R., Ed.; Plenum: New York, NY, USA, 1983; pp. 157–190.

37. Huryn, A.D.; Wallace, J.B. Life history and production of stream insects. *Annu. Rev. Entomol.* **2000**, *45*, 83–110. [CrossRef] [PubMed]

38. McCune, B.; Grace, J.B. *Analysis of Ecological Communities*; MJM Software Design: Gleneden Beach, OR, USA, 2002.

39. Mallis, R.E.; Rieske, L.K. Web orientation and prey resources for web-building spiders in eastern hemlock. *Environ. Entomol.* **2010**, *39*, 1466–1472. [CrossRef] [PubMed]

40. Mallis, R.E.; Rieske, L.K. Arboreal spiders in eastern hemlock. *Environ. Entomol.* **2010**, *40*, 1378–1387. [CrossRef] [PubMed]

41. Adkins, J.K.; Rieske, L.K. Loss of a foundation forest species due to an exotic invader impacts terrestrial arthropod communities. *For. Ecol. Manag.* **2013**, *295*, 126–135. [CrossRef]

42. Adkins, J.K. Impact of the Invasive Hemlock Woolly Adelgid on Headwater Stream Aquatic and Ground-Dwelling Invertebrate Communities. Ph.D. Thesis, University of Kentucky, Lexington, KY, USA, 2012; p. 150.

43. Roberts, S.W.; Tankersley, R.; Orvis, K.H. Assessing the potential impacts to riparian ecosystems resulting from hemlock mortality in Great Smoky Mountains National Park. *Environ. Manag.* **2009**, *44*, 335–345. [CrossRef] [PubMed]

44. Pond, G.J. Patterns of Ephemeroptera taxa loss in Appalachian headwater streams (Kentucky, USA). *Hydrobiologia* **2010**, *641*, 185–201. [CrossRef]

45. Hilsenhoff, W.L. Rapid field assessment of organic pollution with a family-level biotic index. *J. N. Am. Benthol. Soc.* **1988**, *7*, 65–68. [CrossRef]

46. Webster, J.R.; Benfield, E.F. Vascular plant breakdown in freshwater ecosystems. *Annu. Rev. Ecol. Syst.* **1986**, *17*, 567–594. [CrossRef]

47. Cummins, K.W. Structure and function of stream ecosystems. *Bioscience* **1974**, *24*, 631–641. [CrossRef]

48. Cummins, K.W. From headwater streams to rivers. *Am. Biol. Teach.* **1977**, *39*, 305–312. [CrossRef]

49. Short, R.A.; Maslin, P.E. Processing of leaf litter by a stream detritivore: Effect on nutrient availability to collectors. *Ecology* **1977**, *58*, 935–938. [CrossRef]

50. Wallace, J.B.; Webster, J.R. The role of macroinvertebrates in stream ecosystem function. *Annu. Rev. Entomol.* **1996**, *41*, 115–139. [CrossRef] [PubMed]

51. Cummins, K.W.; Petersen, R.C.; Howard, F.O.; Wuycheck, J.C.; Holt, V.I. The utilization of leaf litter by stream detritivores. *Ecology* **1973**, *54*, 336–345. [CrossRef]

52. Heard, S.B. Processing chain ecology: Resource condition and interspecific interactions. *J. Anim. Ecol.* **1994**, *63*, 451–464. [CrossRef]

53. Heard, S.B.; Buchanan, C.K. Grazer-collector facilitation hypothesis supported by laboratory but not field experiments. *Can. J. Fish. Aquat. Sci.* **2004**, *61*, 887–897. [CrossRef]

54. Baxter, C.V.; Fausch, K.D.; Saunders, W.C. Tangled webs: Reciprocal flows of invertebrate prey link streams and riparian zones. *Freshw. Biol.* **2005**, *50*, 201–220. [CrossRef]

55. Lowe, W.H.; Likens, G.E. Moving headwater streams to the head of the class. *Bioscience* **2005**, *55*, 196–197. [CrossRef]

56. MacDonald, L.H.; Coe, D. Influence of headwater streams on downstream reaches in forested areas. *For. Sci.* **2007**, *53*, 148–168.

57. Meyer, J.L.; Strayer, D.L.; Wallace, J.B.; Eggert, S.L.; Helfman, G.S.; Leonard, N.E. The contribution of headwater streams to biodiversity in river networks. *J. Am. Water Resour. Assoc.* **2007**, *43*, 86–103. [CrossRef]

58. Ter Braak, C.J.F.; Verdonschot, P.F.M. Canonical correspondence analysis and related multivariate methods in aquatic ecology. *Aquat. Sci.* **1995**, *57*, 255–289. [CrossRef]

59. McDowell, R.C. *The Geology of Kentucky—A Text to Accompany the Geologic Map of Kentucky*; US Geological Survey Professional Paper 1151-H; Department of the Interior, U.S.G.P.O.: Washington, DC, USA, 1986.

60. Woods, A.J.; Omernik, J.M.; Martin, W.H.; Pond, G.J.; Andrews, W.M.; Call, S.M.; Comstock, J.A.; Taylor, D.D. *Ecoregions of Kentucky (Color Poster with Map, Descriptive Text, Summary Tables, and Photographs)*; US Geological Survey (Map Scale 1:1,000,000): Reston, VA, USA, 2002.

61. Davis, D.H. *The Geography of the Mountains of Eastern Kentucky: A Reconnaissance Study of the Distribution and Activities of Man in that Part of the Cumberland Plateau, Embraced by the Commonwealth*; Series 6; Kentucky Geological Survey: Lexington, KY, USA, 1924.

62. Braun, E.L. *Deciduous Forests of Eastern North America*; Hafner: New York, NY, USA, 1950.

63. Godman, R.; Lancaster, K. Eastern Hemlock. In *Silvics of North America, Volume 1, Conifers*; Burns, R.M., Honkala, B.H., Eds.; USDA Forest Service: Washington, DC, USA, 1990.

64. Wethington, K.; Derting, T.; Kind, T.; Whiteman, H.; Cole, M.; Drew, M. *The Kentucky GAP Analysis Project*; Final Report; Kentucky Department of Fish and Wildlife Resources: Frankfort, KY, USA, 2003.

65. ESRI. *ArcMap*, ESRI: Redlands, CA, USA, 2008.

66. Coleman, T.W.; Clarke, S.R.; Meeker, J.R.; Rieske, L.K. Forest composition following overstory mortality from southern pine beetle and associated treatments. *Can. J. For. Res.* **2008**, *38*, 1406–1418. [CrossRef]

67. United States Department of Agriculture. *Natural Resource Information Service (NRIS): Field Sampled Vegetation (FSVeg) Common Stand Exam, Version 1.5.1*; USDA Forest Service Natural Resource Conservation Service: Washington, DC, USA, 2003.

68. SAS Institute. *SAS 9.3 User's Guide: Statistics, Version 9.3*, SAS Institute: Cary, NC, USA, 2011.

69. Merritt, R.W.; Cummins, K.W. *An Introduction to the Aquatic Insects of North America*, 3rd ed.; Kendall/Hunt Publ. Co.: Dubuque, IA, USA, 1995.

70. Palmer, M.W. Putting things in even better order: The advantages of canonical correspondence analysis. *Ecology* **1993**, *74*, 2215–2230. [CrossRef]

71. Ter Braak, C.J.F. Canonical correspondence analysis: A new eigenvector technique for multivariate direct gradient analysis. *Ecology* **1986**, *67*, 1167–1179. [CrossRef]
72. Elliott, K.J.; Vose, J.M.; Swank, W.T.; Bolstad, P.V. Long-term patterns in vegetation-site relationships in a southern Appalachian forest. *J. Torrey Bot. Soc.* **1999**, *126*, 320–334. [CrossRef]
73. Gabriel, K.R. Biplot display of multivariate matrices for inspection of data and diagnosis. In *Interpreting Multivariate Data*; Barnet, V., Ed.; Wiley-Blackwell: New York, NY, USA, 1981; pp. 571–572.

forests

Article

Foundation Species Loss and Biodiversity of the Herbaceous Layer in New England Forests

Aaron M. Ellison [1,*], Audrey A. Barker Plotkin [1] and Shah Khalid [1,2]

[1]   Harvard Forest, Harvard University, 324 North Main Street, Petersham, MA 01366, USA;
     aabarker@fas.harvard.edu (A.A.B.P.); shahkhalid121@yahoo.com (S.K.)
[2]   Department of Botany, Islamia College, Peshawar 25000, Pakistan
*   Correspondence: aellison@fas.harvard.edu; Tel.: +1-978-756-6178; Fax: +1-978-724-3595

Academic Editor: Diana F. Tomback
Received: 4 November 2015; Accepted: 21 December 2015; Published: 25 December 2015

Abstract: Eastern hemlock (*Tsuga canadensis*) is a foundation species in eastern North American forests. Because eastern hemlock is a foundation species, it often is assumed that the diversity of associated species is high. However, the herbaceous layer of eastern hemlock stands generally is sparse, species-poor, and lacks unique species or floristic assemblages. The rapidly spreading, nonnative hemlock woolly adelgid (*Adelges tusgae*) is causing widespread death of eastern hemlock. Loss of individual hemlock trees or whole stands rapidly leads to increases in species richness and cover of shrubs, herbs, graminoids, ferns, and fern-allies. Naively, one could conclude that the loss of eastern hemlock has a net positive effect on biodiversity. What is lost besides hemlock, however, is landscape-scale variability in the structure and composition of the herbaceous layer. In the Harvard Forest Hemlock Removal Experiment, removal of hemlock by either girdling (simulating adelgid infestation) or logging led to a proliferation of early-successional and disturbance-dependent understory species. In other declining hemlock stands, nonnative plant species expand and homogenize the flora. While local richness increases in former eastern hemlock stands, between-site and regional species diversity will be further diminished as this iconic foundation species of eastern North America succumbs to hemlock woolly adelgid.

Keywords: *Adelges tsugae*; flora; Harvard Forest; herbaceous layer; species diversity; species richness; *Tsuga canadensis*; understory

1. Introduction

Foundation species (*sensu* [1,2]) create and define ecological communities and ecosystems. In general, foundation species are found at the base of food webs [3] and exert bottom-up control on the distribution and abundance of associated biota [4]. Characteristics of foundation species include those of core species [5], dominant species [6], structural species [7], and autogenic ecosystem engineers [8], but none of the latter possesses all of the characteristics of a foundation species [2]. Because of these characteristics, the loss of a foundation species from an ecosystem can have dramatic, cascading effects on other species in the system; on ecosystem stability, resilience, and functioning; and can change our perception of the landscape itself [2].

There is no explicit or implicit magnitude or directionality of the effect of a foundation species on the ecosystem it creates. For example, assemblages of forest understory species in a system dominated by the foundation species *Agathis australis* (D.Don) Lindl. Ex Loudon (Auricariaceae; a.k.a. New Zealand kauri) are different from, but neither more nor less speciose, than forest understory assemblages in systems dominated by other conifers [9]. Alternatively, richness of associated species in an ecosystem defined by a particular foundation species can be greater or less than that of other ecosystems, and the loss of the foundation species could lead either to increased or decreased species richness of the entire associated assemblage [3]. Such variable effects have been found for alpine cushion plants, which have higher alpha (within-cushion) diversity than adjacent, open microhabitats [10–12] but lower beta (between-cushion) diversity [13]. These contrasting effects are thought to arise from creation of "safe sites" for stress-intolerant plants through the local amelioration of stress by cushion plants that simultaneously lead to more homogeneous assemblages on them [13]. In contrast, some perennial kelp species that create complex habitats and provide structure for associated epiphytes have locally negative effects on biodiversity in high stress environments, locally positive effects on biodiversity in less stressful environments, and overall negative effects on biodiversity at larger spatial scales [14].

Eastern hemlock (*Tsuga canadensis* (L.) Càrr.; Pinaceae) is a foundation species of eastern North American forests [2]. The herbaceous understory flora (*sensu* [15]) of hemlock forests is usually thought of as species poor [16–18], but this perception may be due to the much lower understory species richness in the more common second-growth hemlock forests relative to the higher-diversity understory of rare old-growth hemlock forests [19]. However, second-growth hemlock forests dominate the range of the species [20], and the foundation species in these forests is declining and dying rapidly as trees are infested and killed by the nonnative hemlock woolly adelgid (*Adelges tsugae* Annand) [20–22].

In this paper, we describe the effects of loss of eastern hemlock on the local and regional species richness and diversity of the associated forest understory flora. We focus on changes in understory species richness and diversity following experimental removal of the hemlock canopy in central Massachusetts, but place the work in the context of the broad geographic range of eastern hemlock in eastern North America. The impetus for this work came from two directions. First, we were interested in determining whether any rare plant species occur only in the understory of eastern hemlocks or if the loss of eastern hemlock resulted in new habitats for other rare plant species. Second, many of our students and colleagues seeing eastern hemlock forests for the first time are surprised by their species-poor understory. Their implicit expectation is that systems structured by foundation species should be more diverse than systems not structured by foundation species. Thus, we also aim here to provide a more nuanced picture of the interplay between a widespread foundation species and associated diversity at the local and landscape scales.

2. Materials and Methods

2.1. Eastern Hemlock

Tsuga canadensis (eastern hemlock; Pinaceae) is an abundant and widespread late-successional coniferous tree. It grows throughout eastern North America from Georgia north into southern Canada and west into Michigan and Wisconsin [17,20]. In forest stands of the cove forests in the southern part of its range, in mixed forests of New England and southern Canada, and along riparian corridors throughout its >10,000 km^2 range, eastern hemlock can account for >50% of the total basal area [17,23]; see [2,24–26] for detailed discussions of the foundational role of eastern hemlock in stands where it is the dominant species. The forest floor beneath the eastern hemlock canopy is cool and dark [16,27], and the slowly decomposing hemlock needles give rise to a deep organic layer, which is very acidic and low in nutrients [28]. Unique faunal assemblages, including groups of birds [29], arthropods [30–32], and salamanders [33] live in eastern hemlock stands. Fungal diversity in eastern hemlock stands rarely have been studied, but in general is at best equal to, and generally lower than, that in deciduous forests [34–36]. Similarly, both plant diversity and abundance of the species in the herbaceous layer of hemlock understories are low. However, the seed bank and the few established seedlings and saplings respond rapidly to loss of eastern hemlock from disturbances ranging from individual tree falls to wholesale death or removal of entire stands [16,18,25,37–39].

2.2. The Harvard Forest Hemlock Removal Experiment

As part of a long-term, multi-hectare experiment aimed at identifying the effect of loss of eastern hemlock [24], we have documented the response of the herbaceous layer to two different mechanisms of hemlock loss. This Harvard Forest Hemlock Removal Experiment (HF-HeRE) is described in detail in [24]; key details are reiterated here.

HF-HeRE is located in the ≈150-ha Simes Tract (42.47°–42.48° N, 72.22°–72.21° W; 215–300 m above sea level) at the Harvard Forest Long Term Ecological Research Site in Petersham, Massachusetts (complete site description is in [40]). The experiment consists of two blocks, each of which has four ≈ 90 × 90-m (≈0.81-ha) plots. The treatments applied to each plot include: girdling all eastern hemlock individuals (from seedlings to mature trees) to simulate the progressive death-in-place of trees caused by the hemlock woolly adelgid; logging all eastern hemlock individuals ⩾20 cm diameter at breast height (DBH, measured 1.3 m above ground), along with some additional merchantable cordwood (black birch: *Betula lenta* L.; red maple: *Acer rubrum* L.) and sawtimber (red oak: *Quercus rubra* L.; white pine: *Pinus strobus* L.), to simulate a typically intensive level of pre-emptive salvage harvesting; and unmanipulated hemlock controls. Each block also includes a hardwood control dominated by black birch (*Betula lenta* L.) and red maple (*Acer rubrum* L.) that represents the young stands expected to replace eastern hemlock as it is lost from the forests of northeastern North America [41].

When HF-HeRE was established in 2003, the hemlock woolly adelgid had not yet colonized the forest interior at Harvard Forest. As we expected it to eventually colonize our hemlock control plots, HF-HeRE was designed explicitly to contrast the effects on these forests from physically disintegrating trees (resulting from girding them) with removal of them from the site (following logging). Since the adelgid colonized the hemlock controls—which occurred in 2009 and 2010—we have been able to contrast the effects of physical disintegration of eastern hemlock (in the girdled plots) with the effects of the adelgid (in the hemlock control plots), which includes not only physical disintegration but also changes caused by the adelgid directly, including, e.g., nitrogen inputs [42,43].

2.3. The Herbaceous Layer

In 2003, prior to canopy manipulations, we established two transects running through the central 30 × 30 m of each canopy manipulation plot and the associated hemlock and hardwood control plots to quantify understory richness, cover, and density. Five 1-m^2 subplots were spaced evenly along each transect and have been sampled annually since 2003. In each subplot, percent cover of herbs, shrubs, ferns and grasses was estimated to the nearest one percent. Grasses and sedges were identified only to genus as most lacked flowers or fruits necessary for accurate species-level identification. Each year, we also noted all understory species occurring in the entire central 30 × 30-m area of each plot; these incidence-level data encompass not only relatively common species enumerated along our sample transects but also the more uncommon species. Nomenclature follows [44]. Data reported herein were collected at HF-HeRE from 2003 through 2014.

We estimated species diversity in the different treatments from the incidence data (*i.e.*, presence-absence data in the entire 30 × 30-m central area of each plot) using Hill numbers [45]; comparisons among plots used previously published methods [46,47]. Changes in similarity and composition of herbaceous assemblages were examined using principal components analysis on centered and standardized percent cover data (collected along the two transects within each plot). All analyses were done using the R statistical software system [48], version 3.2.2 and routines within the *SpadeR* package [49] for diversity calculations, prcomp within the *stats* library for principal components analysis; and aov and TukeyHSD within the *stats* library for repeated-measures (random-effects) analysis of variance. Data and code are available from the Harvard Forest Data Archive, dataset HF106 [50].

3. Results and Discussion

3.1. The Response of the Herbaceous Layer to Experimental Removal of Eastern Hemlock

In total and across all years, we found 73 shrub, herb, graminoid, and fern/fern-ally species growing in the eight experimental plots (complete list of species is in [50]). Observed species richness in the herbaceous layer (pooled across all years) was highest in the girdled and hardwood control plots, and lowest in the hemlock control plots (Table 1). Estimated diversity (as Hill numbers of richness, Shannon diversity, and Simpson's diversity) was significantly lower in the hemlock controls than in the other three canopy-manipulation treatments (Table 1). Mean pairwise similarity between canopy manipulation treatments of the herbaceous layer averaged 0.55 (Figure 1). The herbaceous layers of the girdled and logged treatments were most similar (Jost's D = 0.824), whereas, they were the least similar between the hemlock control and the logged plots (D = 0.383). Although the number of shared species (bracketed numbers in Figure 1) varied four-fold among pairs of treatments, pairwise similarities were related neither to the total number of understory species (Table 1) nor to the numbers of shared species (Figure 1).

Table 1. Species richness (all years pooled) in the Harvard Forest Hemlock Removal Experiment. Values given are number of incidences (occurrences of species in both plots of each canopy manipulation treatment), number of species observed (S_{obs}), and estimated Hill numbers (95% confidence intervals) for species richness (^0q), Shannon diversity (^1q), and Simpson's diversity (^2q).

	Incidences	S_{obs}	^0q	^1q	^2q
Hemlock control	188	18	19.9 (16.9, 23.0)	15.9 (14.8, 17.1)	14.9 (13.4, 16.8)
Girdled	388	53	54.2 (45.0, 63.5)	42.8 (39.5, 46.0)	36.7 (33.3, 40.0)
Logged	305	38	50.2 (17.7, 82.8)	30.4 (27.2, 33.5)	25.9 (23.1, 28.7)
Hardwood control	616	51	52.6 (40.0, 65.2)	41.1 (39.1, 43.1)	37.2 (35.0, 39.4)
All treatments pooled	1497	73	73.5 (73.0, 81.4)	53.8 (50.2, 54.9)	43.2 (39.3, 44.6)

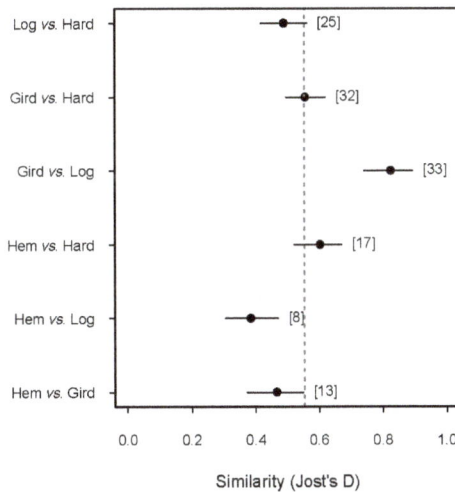

Figure 1. Pairwise similarities in understory species composition (with 95% confidence intervals) between the canopy manipulation treatments in the Harvard Forest Hemlock Removal Experiment. Abbreviations for treatments are: Hem—Hemlock controls; Gird—Hemlocks girdled; Log—Hemlocks cut and removed; Hard—Hardwood controls. Numbers in brackets are the total number of shared species for each pairwise comparison, and the grey dashed line is the average overall pairwise similarity.

The vegetation composition in the herbaceous layer varied substantially among the four canopy manipulation treatments and through time (Figure 2). The first three principal axes accounted for 50% of the variance in vegetation composition, and visual examination of the scree plot suggested that subsequent axes were uninformative. Of the 49 shrub, herb, graminoid, fern, and fern-ally species identified in the sampled transects across all plots, only 18 loaded heavily on the first three principal axes (absolute value of their loading ⩾0.25), and these "important" taxa segregated cleanly among them (Table 2). In Figure 2, principal axis 1 emphasizes understory herbs that are common in all plots, including hemlock-dominated ones, whereas principal axes 2 and 3 have taxa that are more common in early- to mid-successional mixed forests stands (e.g., *Epigaea repens, Rhododendron periclymenoides, Dyropteris carthusiana,* and *Carex* spp.) or commonly recruit after disturbance (such as the two *Rubus* species, *Panicum* sp., and the nonnative *Berberis thunbergii*) (Table 2). No state-listed rare, threatened, or endangered taxa were found in any of the plots.

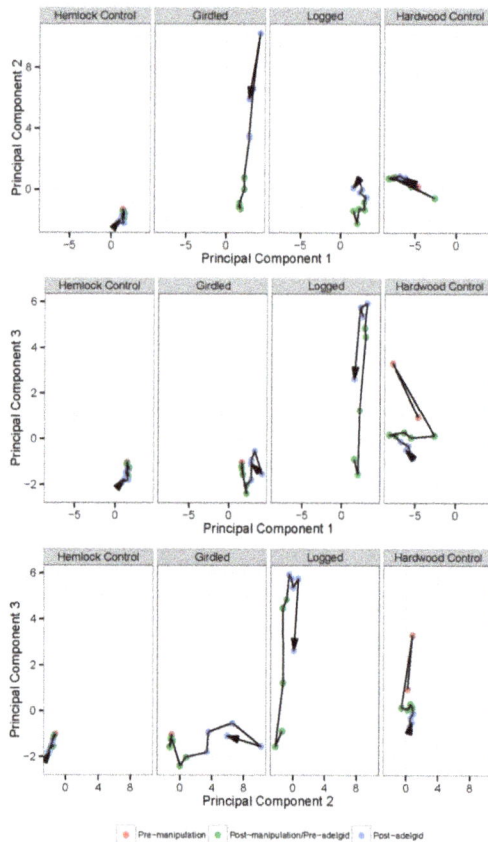

Figure 2. Temporal trajectories of understory vegetation composition in principal components space in the Harvard Forest Hemlock Removal Experiment. Each point represents the vegetation composition in a given year, and the line traces the trajectory through time from before canopy manipulations (2003–2004: red points), after manipulations but prior to adelgid colonization of the plots (2005–2009: green points), and after the adelgid had colonized the plots (2010–2014: blue points). In four of the panels (PC1 *vs.* PC2 for the girdled plots, and all three panels for the logged plots), the 2003–2004 vegetation composition was not distinguishable from the 2005 data and so the red symbols are not visible.

Table 2. Loadings of species on each of the three principal components axes illustrated in Figure 2. Only species for which the absolute value of the loading ⩾0.25 are shown.

Species	PC-1	PC-2	PC-3
Aralia nudicaulis L.	−0.26		
Dendrolycopodium obscurum (L). A. Haines	−0.26		
Lysimachia borealis (Raf.) U. Manns & A. Anderb.	−0.25		
Berberis thunbergii DC.		0.36	
Lonicera canadensis Bartr. Ex Marsh.		0.36	
Dryopteris carthusiana (Vill.) H.P. Fuchs		0.32	
Ilex verticillata (L.) Gray		0.32	
Osmundastrum cinnamomeum (L.) C. Presl		0.31	
Epigaea repens L.		0.31	
Viburnum acerifolium L.		0.31	
Rhododendron periclymenoides (Michx.) Shinners		0.26	
Lysimachia quadrifolia L.			0.38
Carex sp.			0.34
Rubusallegheniensis Porter			0.31
Araliahispida Veng.			0.30
Rubusidaeus L.			0.28
Carex cf. *pennsylvanica* Lam.			0.28
Panicum sp.			0.27

The overall impression from Figure 2 is that the understory of the hemlock control plots has been relatively stable through time, a result that is attributable primarily to its low species richness and the sparseness of what is there. We expect this to change as the overstory declines in coming years. The young hardwood control stands, in contrast, exhibit more temporal changes, both because the understory has more species and more cover to begin with, and because of the interaction of canopy closure as the trees grow and new gap creation from treefalls.

Repeated-measures analysis of variance (with pre-manipulation, post-manipulation/pre-adelgid infestation, and post-adelgid as the repeating groups or "temporal strata") revealed that principal axis 1 scores differed among the canopy manipulation treatments (Table 3). The first principal axis scores of the hardwood controls were significantly lower ($p \leqslant 0.05$, all pairwise comparisons using Tukey's HSD test) than the other three canopy manipulation treatments (Hemlock Control = Girdled = Logged < Hardwood Control) and these comparisons did not change after treatments were applied or the adelgid colonized the plot (Table 3). In contrast, principal axis 2 scores were significantly different for all four canopy manipulation treatments (Girdled > Hardwood Control > Hemlock Control = Logged), and differed significantly after the adelgid had colonized the plots relative to the preceding 7 years (Pre-treatment = Post-treatment/pre-adelgid < Post-adelgid). Finally, there were significant differences among canopy manipulation treatments for principal axis 3 (Logged > Hardwood Control > Girdled = Hemlock Control). The differences in PC-2 and PC-3 between canopy manipulation treatments were most dramatic in pairwise comparisons between treatments after the adelgid had colonized the plots relative to the previous two temporal strata (Figures 3 and 4; nested terms in Table 3).

These experimental results support earlier results that showed a rapid increase in seedling germination [38,39], density [18,25] and diversity [18,25,37] as hemlock dies and light levels at the forest floor increase [27]. Unexpectedly, however, the composition of the herbaceous layer shifted further after the adelgid colonized the plots, albeit unevenly across treatments (*i.e.*, the significant term representing canopy manipulation treatment nested within temporal stratum). The difference in interaction terms observed in PC-2 (Figure 3) most likely reflects a rapid decline in light levels of the understory in the girdled plots as birch saplings are growing exponentially [25]. This growth is occurring concomitantly with adelgid infestation in the hemlock controls. Although we have observed steady increases in light in the hemlock control plots as the adelgid increased in abundance [51], this is not yet strongly affecting understory vegetation in the hemlock controls. We expect that in coming years, additional nonlinear changes in soil N in the hemlock control plots as a function of the adelgid, as suggested by [42,43] and observed by [51] may affect understory composition. Soil N initially increases because carbon sloughing off from the waxy coating of the adelgid provides additional energy for microbial N immobilization of the relatively N-rich needles of infested trees [42]. Consequently, Nitrogen fluxes initially decrease with infestation, but later rise as hemlock declines and is replaced by deciduous trees, whose litter has a higher percentage of N [42]. The differences in interaction terms for PC-3 (Figure 4) suggest increased rate of succession following logging, as the small hemlocks that were not removed during the logging operation have not yet succumbed to the adelgid.

Table 3. Results of repeated-measures ANOVA on the first three principal axis scores of vegetation composition in the Harvard Forest Hemlock Removal Experiment. Factors include three temporal strata (pre-manipulation (2003–2004); post-manipulation but prior to adelgid colonization (2005–2009); and after the adelgid had colonized the plots (2010–2014)) and four canopy manipulation treatments (hemlock control, girdled, logged, and hardwood control). To account for the repeated measures, canopy manipulation treatments are nested within temporal strata.

Response	Factor	df	MS	F	P
PC-1	Temporal stratum	2	1.05	1.21	0.31
	Treatment	3	196.21	226.23	<0.001
	Treatment within stratum	6	0.76	0.88	0.52
	Residual	36	0.87		
PC-2	Temporal stratum	2	22.95	21.68	<0.001
	Treatment	3	33.31	31.47	<0.001
	Treatment within stratum	6	14.87	14.05	<0.001
	Residual	36	1.06		
PC-3	Temporal stratum	2	3.40	2.53	0.09
	Treatment	3	44.97	33.44	<0.001
	Treatment within stratum	6	10.66	7.93	<0.001
	Residual	36	1.34		

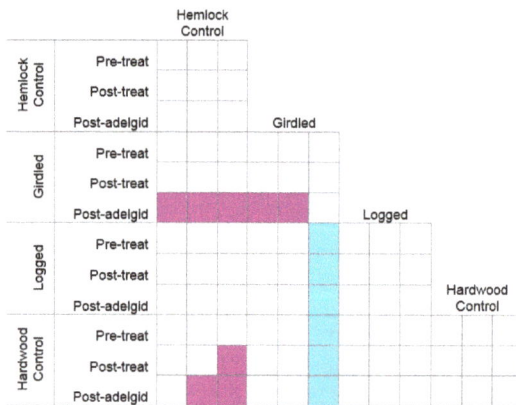

Figure 3. *Post-hoc*, pair-wise comparisons (Tukey's Honest Significant Difference Test) of the response (principal axis 2) to canopy manipulation treatments (hemlock control, girdled, logged, and hardwood control) repeated within temporal strata (pre-manipulation (2003–2004); post-manipulation but prior to adelgid colonization (2005–2009); and after the adelgid had colonized the plots (2010–2014)). For each pair-wise comparison, differences between PC-2 of a given row and PC-2 of a given column are significantly greater than (magenta) or less than (blue) zero ($p \leqslant 0.05$).

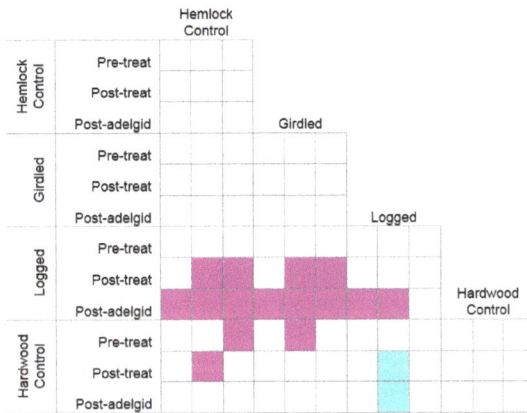

Figure 4. *Post-hoc*, pair-wise comparisons (Tukey's Honest Significant Difference Test) of the response (principal axis 3) to canopy manipulation treatments (hemlock control, girdled, logged, and hardwood control) repeated within temporal strata (pre-manipulation (2003–2004); post-manipulation but prior to adelgid colonization (2005–2009); and after the adelgid had colonized the plots (2010–2014)). For each pair-wise comparison, differences between PC-3 of a given row and PC-3 of a given column are significantly greater than (magenta) or less than (blue) zero ($p \leqslant 0.05$).

3.2. The Herbaceous Layer in Other Eastern Hemlock Stands

Substantial differences in the herbaceous flora of eastern hemlock stands relative to nearby stands dominated by other conifers or mixed hardwood have been documented repeatedly [16–19,52–54]. In Indiana, Daubenmire placed 25 1-m^2 quadrats in hemlock forest and found only one species of herb (*Monotropa uniflora* L.), no shrubs, and three trees [16]. In nearby beech-maple forests, however, he found 14 species of herbs, four shrubs, and seven trees. He also noted that *Monotropa* growing in the forest understory "was as dominant in the Hemlock as Sugar Maple was in the Beech-Maple" [16]. Likewise, under hemlocks growing in southeast Ohio, the ground-layer was >80% bare ground, and herbaceous species richness was correspondingly low [54].

Although the herbaceous flora in hemlock stands of southeast North America is much richer than that in the northeastern part of hemlock's range, it also is species poor relative to nearby hardwood stands. Unlike in the northeast, however, the herbaceous layer of southeast hemlock forests has been found repeatedly to be distinctive [52,53,55]. In the southeast, *Rhododendron maximum* L. can grow in dense thickets, and the shrub *Leucothoe fontanesiana* (Steud.) Sleumer, the ground-creeping *Mitchella repens* L. and the herb *Hexastylis shuttleworthii* (Britten and Baker f.) Small are predictably associated with hemlock stands [17,20].

Prior to the arrival in the 1980s of the hemlock woolly adelgid in northeast North America [21], intact hemlock stands from Wisconsin (USA) to Nova Scotia (Canada) had an average percent cover in the herbaceous layer of ≈5% and consisted primarily of common species [17]. Furthermore, these stands had no compositionally distinctive groups of understory species. Rather, all herbaceous-layer species growing in these hemlock stands also were found in northern mixed hardwood stands [17]. In contrast, many herbaceous species common in mixed hardwood stands were rare in hemlock-dominated stands [17].

When individual eastern hemlock trees die, fall, and form gaps, species richness of the herbaceous layer increases rapidly [16,18,34]. Early studies of the herbaceous flora of hemlock forests were done in what were thought to be "climax" [16], "old-growth" [18] or "virgin" [52,53] stands. Gap creation in old-growth stands in Wisconsin led to increases in richness and cover of herbaceous species in gaps relative to intact forest [18]. This result was echoed by those of D'Amato *et al.* [19] in relict old-growth stands in Massachusetts (USA). Although these old-growth hemlock stands supported twice the number of understory species and had four times the understory vegetation cover as nearby second-growth hemlock stands [19], these differences were attributable to small-scale disturbances (treefall gaps, self-thinning), leading to lower density canopies and more light reaching the forest floor. Such gap dynamics also alter understory diversity in hardwood-dominated stands (Figure 2).

Similar responses occur when entire stands of hemlock are killed by the hemlock woolly adelgid. For example, between 1994 and 2006, up to 42% of individuals died from adelgid infestation in hemlock stands in the Delaware Water Gap National Recreation Area [56]. During this interval, species richness and cover of bryophytes (mosses and liverworts) more than doubled, in large part because of the amount of new habitat (dead and fallen branches and boles) and higher light availability on streambanks. In the same plots, all understory ferns, herbs (except for *Streptopus amplexifolius* (L.) DC), shrubs, and trees increased in occurrence and abundance as hemlock declined [57]. At the same time, there was a large increase—further accelerated by increasing density of deer—in abundance of nonnative species, including *Ailanthus altissima* (P.Mill.) Swingle, *Alliaria petiolata* (M. Bieb) Cavara and Grande, *Berberis thunbergii*, *Microstegium vimineum* (Trin.) A.Camus, and *Rosa multiflora* Thunb. ex Murray [58].

4. Conclusions

In sum, observational and experimental data illustrate that the herbaceous layer of eastern hemlock stands generally is sparse and species-poor. With few exceptions, hemlock stands have not been found to have a unique set of associated understory species. Rather, species that are common in the hemlock understory also are common in mixed hardwood stands, whereas other species that are common in mixed hardwoods are uncommon in hemlock stands. Death of individual hemlock trees or of whole stands routinely leads to rapid increases in species richness and percent cover of shrubs, herbs, graminoids, ferns, and fern-allies.

What is lost besides hemlock, however, is variability in the structure and composition of the herbaceous layer. In the Harvard Forest Hemlock Removal Experiment, initial removal of hemlock by either girdling or logging, and subsequently by the hemlock woolly adelgid led to a proliferation of early-successional and disturbance-dependent understory species. In other declining hemlock stands, nonnative species expand and homogenize the flora. While local richness increases, between-site and regional species diversity can be expected to decline further as eastern hemlock—an iconic foundation species of eastern North America—is lost from many forest stands.

Acknowledgments: We thank Diana Tomback for inviting us to prepare this contribution. A.M.E. and A.A.B.P.'s work on hemlock forests is part of the Harvard Forest Long Term Ecological Research (LTER) program, supported by grants 06-20443 and 12-37491 from the US National Science Foundation. S.K. was supported by a scholarship from the Higher Education Commission of Pakistan. Ally Degrassi pointed us to useful background references, and Simon Smart and two anonymous reviewers provided helpful comments on the initial submission. This paper is a contribution of the Harvard Forest LTER program.

Author Contributions: A.M.E. and A.A.B.P. conceived of the study, did data analysis, and co-wrote the paper. A.A.B.P. and S.K. collected and organized the data, and co-wrote the paper.

Conflicts of Interest: The authors declare no conflict of interest.

References

1. Dayton, P.K. Toward an understanding of community resilience and the potential effects of enrichments to the benthos at McMurdo Sound, Antarctica. In Proceedings of the Colloquium on Conservation Problems in Antarctica, Blacksburg, VA, USA, 10–12 September 1971; Parker, B.C., Ed.; Allen Press: Lawrence, KS, USA, 1972; pp. 81–95.

2. Ellison, A.M.; Bank, M.S.; Clinton, B.D.; Colburn, E.A.; Elliott, K.; Ford, C.R.; Foster, D.R.; Kloeppel, B.D.; Knoepp, J.D.; Lovett, G.M.; *et al.* Loss of foundation species: Consequences for the structure and dynamics of forested ecosystems. *Front. Ecol. Environ.* **2005**, *9*, 479–486. [CrossRef]

3. Baiser, B.; Whitaker, N.; Ellison, A.M. Modeling foundation species in food webs. *Ecosphere* **2013**, *4*, 146. [CrossRef]

4. Sackett, T.E.; Record, S.; Bewick, S.; Baiser, B.; Sanders, N.J.; Ellison, A.M. Response of macroarthropod assemblages to the loss of hemlock (*Tsuga canadensis*), a foundation species. *Ecosphere* **2011**, *2*, 74. [CrossRef]

5. Hanski, I. Dynamics of regional distribution: The core and satellite species hypothesis. *Oikos* **1982**, *38*, 210–221. [CrossRef]

6. Grime, J.P. Dominant and subordinate components of plant communities: Implications for succession, stability and diversity. In *Colonization, Succession and Stability*; Gray, A.J., Crawley, M.J., Eds.; Blackwell Scientific Publishers: Oxford, UK, 1984; pp. 413–428.

7. Huston, M.A. *Biological Diversity: The Coexistence of Species on Changing Landscapes*; Cambridge University Press: Cambridge, UK, 1994.

8. Jones, C.G.; Lawton, J.H.; Shachak, M. Organisms as ecosystem engineers. *Oikos* **1994**, *69*, 373–386. [CrossRef]

9. Wyse, S.V.; Burns, B.R.; Wright, S.D. Distinctive vegetation communities are associated with the long-lived conifer *Agathis australis* (New Zealand kauri, Araucariaceae) in New Zealand rainforests. *Austral Ecol.* **2014**, *39*, 388–400. [CrossRef]

10. Schöb, C.; Butterfield, B.J.; Pugnaire, F.I. Foundation species influence trait-based community assembly. *New Phytol.* **2012**, *196*, 824–834. [CrossRef] [PubMed]

11. Butterfield, B.J.; Cavieres, L.A.; Callaway, R.M.; Cook, B.J.; Kikvidze, Z.; Lortie, C.J.; Michalet, R.; Pugnaire, F.I.; Schöb, C.; Xiao, S.; *et al.* Alpine cushion plants inhibit the loss of phylogenetic diversity in severe environments. *Ecol. Lett.* **2013**, *16*, 478–486. [CrossRef] [PubMed]

12. Cavieres, L.A.; Brooker, R.W.; Butterfield, B.J.; Cook, B.J.; Kikvidze, Z.; Lortie, C.J.; Michalet, R.; Pugnaire, F.I.; Schöb, C.; Xiao, S.; *et al.* Facilitative plant interactions and climate simultaneously drive alpine plant diversity. *Ecol. Lett.* **2014**, *17*, 193–202. [CrossRef] [PubMed]

13. Kikvidze, Z.; Brooker, R.W.; Buttefield, B.J.; Callaway, R.M.; Cavieres, L.A.; Cook, B.J.; Lortie, C.J.; Michalet, R.; Pugnaire, F.I.; Xiao, S.; *et al.* The effects of foundation species on community assembly: A global study of alpine cushion plant communities. *Ecology* **2015**, *96*, 2064–2069. [CrossRef] [PubMed]

14. Hughes, B.B. Variable effects of a kelp foundation species on rocky intertidal diversity and species interactions in central California. *J. Exp. Mar. Biol. Ecol.* **2010**, *393*, 90–99. [CrossRef]

15. Gilliam, F.S. The herbaceous layer—The forest between the trees. In *The Herbaceous Layer in Forests of Eastern North America*, 2nd ed.; Gilliam, F.S., Ed.; Oxford University Press: Oxford, UK, 2014; pp. 1–12.

16. Daubenmire, R.F. The relation of certain ecological factors to the inhibition of forest floor herbs under hemlock. *Butl. Univ. Bot. Stud.* **1929**, *1*, 61–76. Available online: http://digitalcommons.butler.edu/botanical/vol1/iss1/7 (accessed on 3 November 2015).

17. Rogers, R.S. Hemlock stands from Wisconsin to Nova Scotia: Transitions in understory composition along a floristic gradient. *Ecology* **1980**, *61*, 178–193. [CrossRef]

18. Mladenoff, D.J. The relationship of the soil seed bank and understory vegetation in old-growth northern hardwood-hemlock treefall gaps. *Can. J. Bot.* **1990**, *68*, 2714–2721. [CrossRef]

19. D'Amato, A.W.; Orwig, D.A.; Foster, D.R. Understory vegetation in old-growth and second-growth *Tsuga canadensis* forests in western Massachusetts. *For. Ecol. Manag.* **2009**, *257*, 1043–1052. [CrossRef]

20. Abella, S.R. Imacts and management of hemlock woolly adelgid in national parks of the eastern United States. *Southeast. Nat.* **2014**, *13*, 16–45.

21. McClure, M. Biology and control of hemlock woolly adelgid. *Bull. Conn. Agric. Exp. Stn* **1987**, *851*, 1–9.

22. Orwig, D.A.; Foster, D.R. Forest response to the introduced hemlock woolly adelgid in southern New England, USA. *J. Torrey Bot. Soc.* **1998**, *125*, 60–73. [CrossRef]

23. Smith, W.B.; Miles, P.D.; Perry, C.H.; Pugh, S.A. *Forest Resources of the United States, 2007*; General Technical Report WO-78; USDA Forest Service: Washington, DC, USA, 2009.

24. Ellison, A.M.; Barker Plotkin, A.A.; Foster, D.R.; Orwig, D.A. Experimentally testing the role of foundation species in forests: The Harvard Forest Hemlock Removal Experiment. *Methods Ecol. Evol.* **2010**, *1*, 168–179. [CrossRef]

25. Orwig, D.A.; Barker Plotkin, A.A.; Davidson, E.A.; Lux, H.; Savage, K.E.; Ellison, A.M. Foundation species loss affects vegetation structure more than ecosystem function in a northeastern USA forest. *PeerJ* **2013**, *1*, e41. [CrossRef] [PubMed]

26. Ellison, A.M. Reprise: Eastern hemlock as a foundation species. In *Hemlock: A Forest Giant on the Edge*; Foster, D.R., Ed.; Yale University Press: New Haven, CT, USA, 2014; pp. 165–171.

27. Lustenhouwer, M.N.; Nicoll, L.; Ellison, A.M. Microclimatic effects of the loss of a foundation species from New England forests. *Ecosphere* **2012**, *3*, 26. [CrossRef]

28. Cobb, R.C.; Orwig, D.A.; Currie, S.J. Decomposition of green foliage in eastern hemlock forests of southern New England impacted by hemlock woolly adelgid infestations. *Can. J. For. Res.* **2006**, *36*, 1–11. [CrossRef]

29. Tingley, M.W.; Orwig, D.A.; Field, R.; Motzkin, G. Avian response to removal of a forest dominant: Consequences of hemlock woolly adelgid infestations. *J. Biogeogr.* **2002**, *29*, 1505–1516. [CrossRef]

30. Ellison, A.M.; Chen, J.; Díaz, D.; Kammerer-Burnham, C.; Lau, M. Changes in ant community structure and composition associated with hemlock decline in New England. In Proceedings of the 3rd Symposium on Hemlock Woolly Adelgid in the Eastern United States, Asheville, NC, USA, 1–3 February 2005; Onken, B., Reardon, R., Eds.; US Department of Agriculture, US Forest Service Forest Health Technology Enterprise Team: Morgantown, WV, USA, 2005; pp. 280–289.

31. Dilling, C.; Lambdin, P.; Grant, J.; Buck, L. Insect guild structure associated with eastern hemlock in the southern Appalachians. *Environ. Entomol.* **2007**, *36*, 1408–1414. [CrossRef]

32. Rohr, J.R.; Mahan, C.G.; Kim, K. Response of arthropod biodiversity to foundation species declines: The case of the eastern hemlock. *For. Ecol. Manag.* **2009**, *258*, 1503–1510. [CrossRef]

33. Mathewson, B.G. Eastern red-backed salamander relative abundance in eastern hemlock-dominated and mixed deciduous forests at the Harvard Forest. *Northeast. Nat.* **2009**, *16*, 1–12. [CrossRef]

34. Porter, T.M.; Skillman, J.E.; Moncalvo, J.M. Fruiting body and soil rDNA sampling detects complementary assemblage of Agaricomycotina (Basidiomycota, Fungi) in a hemlock-dominated forest plot in southern Ontario. *Mol. Ecol.* **2008**, *17*, 3037–3050. [CrossRef] [PubMed]

35. Baird, R.E.; Watson, C.E.; Woolfok, S. Microfungi from bark of healthy and damaged American beech, Fraser fir, and eastern hemlock trees during an all taxa biodiversity inventory in forests of the Great Smoky Mountains National Park. *Southeast. Nat.* **2007**, *6*, 67–82. [CrossRef]

36. Baird, R.E.; Woolfok, S.; Watson, C.E. Microfungi of forest litter from healthy American beech, Fraser fir, and eastern hemlock stands in Great Smoky Mountains National Park. *Southeast. Nat.* **2009**, *8*, 609–630. [CrossRef]

37. Catovsky, S.; Bazzaz, F. The role of resource interactions and seedling regeneration in maintaining a positive feedback in hemlock stands. *J. Ecol.* **2000**, *88*, 100–112. [CrossRef]

38. Sullivan, K.A.; Ellison, A.M. The seed bank of hemlock forests: Implications for forest regeneration following hemlock decline. *J. Torrey Bot. Soc.* **2006**, *133*, 393–402. [CrossRef]

39. Farnsworth, E.J.; Barker Plotkin, A.A.; Ellison, A.M. The relative contributions of seed bank, seed rain, and understory vegetation dynamics to the reorganization of *Tsuga canadensis* forests after loss due to logging or simulated attack by *Adelges tsugae*. *Can. J. For. Res.* **2012**, *42*, 2090–2105. [CrossRef]

40. Ellison, A.M.; Lavine, M.; Kerson, P.B.; Barker Plotkin, A.A.; Orwig, D.A. Building a foundation: Land-use history and dendrochronology reveal temporal dynamics of a *Tsuga canadensis* (Pinaceae) forest. *Rhodora* **2014**, *116*, 377–427. [CrossRef]

41. Albani, M.; Moorcroft, P.R.; Ellison, A.M.; Orwig, D.A.; Foster, D.R. Predicting the impact of hemlock woolly adelgid on carbon dynamics of eastern US forests. *Can. J. For. Res.* **2010**, *40*, 119–133. [CrossRef]

42. Stadler, B.; Müller, T.; Orwig, D. The ecology of energy and nutrient fluxes in hemlock forests invaded by hemlock woolly adelgid. *Ecology* **2006**, *87*, 1792–1804. [CrossRef]

43. Stadler, B.; Müller, T.; Orwig, D.; Cobb, R. Hemlock woolly adelgid in New England forests: Transforming canopy impacts transforming ecosystem processes and landscapes. *Ecosystems* **2005**, *8*, 233–247. [CrossRef]

44. Haines, A. *Flora Novae Angliae: A Manual for the Identification of Native and Naturalized Vascular Plants of New England*; Yale University Press: New Haven, CT, USA, 2011.

45. Chao, A.; Jost, L. Estimating diversity and entropy profiles via discovery rates of new species. *Methods Ecol. Evol.* **2015**, *6*, 873–882. [CrossRef]

46. Chao, A.; Jost, L.; Chiang, S.C.; Jiang, Y.H.; Chazdon, R. A two-stage probabilistic approach to multiple-community similarity indices. *Biometrics* **2008**, *64*, 1178–1186. [CrossRef] [PubMed]

47. Jost, L. GST and its relatives do not measure differentiation. *Mol. Ecol.* **2008**, *17*, 4015–4026. [CrossRef] [PubMed]

48. R Core Team. *R: A Language and Environment for Statistical Computing*; Version 3.2.2; R Foundation for Statistical Computing: Vienna, Austria, 2015; Available online: http://r-project.org/ (accessed on 1 July 2015).

49. Chao, A.; Ma, K.H.; Hsieh, T.C. SpadeR: Species Prediction and Diversity Estimation with R, Version 0.1.0. Available online: http://chao.stat.nthu.edu.tw/blog/software-download/ (accessed on 18 October 2015).

50. Ellison, A.; Plotkin, B.A. Understory Vegetation in Hemlock Removal Experiment at Harvard Forest Since 2003. Harvard Forest Data Archive: HF106. 2005. Available online: http://harvardforest.fas. harvard.edu:8080/exist/apps/datasets/showData.html?id=hf106 (accessed on 3 November 2015).

51. Kendrick, J.A.; Ribbons, R.R.; Classen, A.T.; Ellison, A.M. Changes in canopy structure and ant assemblages affect soil ecosystem variables as a foundation species declines. *Ecosphere* **2015**, *6*, 77. [CrossRef]

52. Oosting, H.J.; Bordeau, P.F. Virgin hemlock forest segregates in the Joyce Kilmer Memorial Forest of western North Carolina. *Bot. Gaz.* **1955**, *116*, 340–359. [CrossRef]

53. Bieri, R.; Anliot, S.F. The structure and floristic composition of a virgin hemlock forest in West Virginia. *Castanea* **1965**, *30*, 205–226.

54. Martin, K.L.; Goebel, P.C. The foundation species influence of eastern hemlock (*Tsuga canadensis*) on biodiversity and ecosystem function on the unglaciated Allegheny Plateau. *For. Ecol. Manag.* **2013**, *289*, 143–152. [CrossRef]

55. Abella, S.R.; Shelburne, V.B. Ecological species groups of South Carolina's Jocassee Gorges, Southern Applachian Mountains. *J. Torrey Bot. Soc.* **2004**, *131*, 200–222. [CrossRef]

56. Cleavitt, N.L.; Eschtruth, A.K.; Battles, J.J.; Fahey, T.J. Bryophyte response to eastern hemlock decline casued by hemlock woolly adelgid infestation. *J. Torrey Bot. Soc.* **2008**, *135*, 12–25. [CrossRef]

57. Eschtruth, A.K.; Cleavitt, N.L.; Battles, J.J.; Evans, R.A.; Fahey, T.J. Vegetation dynamics in declining eastern hemlock stands: 9 years of forest response to hemlock woolly adelgid infestation. *Can. J. For. Res.* **2006**, *36*, 1435–1450. [CrossRef]

58. Eschtruth, A.K.; Battles, J.J. Acceleration of exotic plant invasion in a forested ecosystem by a generalist herbivore. *Conserv. Biol.* **2009**, *23*, 388–399. [CrossRef] [PubMed]

forests

MDPI

Article

Community Structure, Biodiversity, and Ecosystem Services in Treeline Whitebark Pine Communities: Potential Impacts from a Non-Native Pathogen

Diana F. Tomback [1,*], Lynn M. Resler [2], Robert E. Keane [3], Elizabeth R. Pansing [1], Andrew J. Andrade [1] and Aaron C. Wagner [1]

[1] Department of Integrative Biology, CB 171, University of Colorado Denver, P.O. Box 173364, Denver, CO 80217, USA; elizabeth.pansing@ucdenver.edu (E.R.P.); andrew.andrade@ucdenver.edu (A.J.A.); aaron.wagner@ucdenver.edu (A.C.W.)

[2] Department of Geography, Virginia Tech, 115 Major Williams Hall, Blacksburg, VA 24061, USA; resler@vt.edu

[3] U.S. Forest Service, Rocky Mountain Research Station, Missoula Fire Sciences Laboratory, 5775 US Hwy 10 West, Missoula, MT 59808, USA; rkeane@fs.fed.us

* Correspondence: diana.tomback@ucdenver.edu; Tel.: +1-303-556-2657; Fax: +1-303-556-4352

Academic Editor: Eric J. Jokela
Received: 11 November 2015; Accepted: 6 January 2016; Published: 19 January 2016

Abstract: Whitebark pine (*Pinus albicaulis*) has the largest and most northerly distribution of any white pine (Subgenus *Strobus*) in North America, encompassing 18° latitude and 21° longitude in western mountains. Within this broad range, however, whitebark pine occurs within a narrow elevational zone, including upper subalpine and treeline forests, and functions generally as an important keystone and foundation species. In the Rocky Mountains, whitebark pine facilitates the development of krummholz conifer communities in the alpine-treeline ecotone (ATE), and thus potentially provides capacity for critical ecosystem services such as snow retention and soil stabilization. The invasive, exotic pathogen *Cronartium ribicola*, which causes white pine blister rust, now occurs nearly rangewide in whitebark pine communities, to their northern limits. Here, we synthesize data from 10 studies to document geographic variation in structure, conifer species, and understory plants in whitebark pine treeline communities, and examine the potential role of these communities in snow retention and regulating downstream flows. Whitebark pine mortality is predicted to alter treeline community composition, structure, and function. Whitebark pine losses in the ATE may also alter response to climate warming. Efforts to restore whitebark pine have thus far been limited to subalpine communities, particularly through planting seedlings with potential blister rust resistance. We discuss whether restoration strategies might be appropriate for treeline communities.

Keywords: *Pinus albicaulis*; alpine-treeline ecotone; biodiversity; community structure; keystone species; foundation species; ecosystem services; exotic pathogen; white pine blister rust; restoration

1. Introduction

Temperate zone coniferous forests are important contributors to global annual net primary productivity, standing biomass, and biodiversity ([1] and references therein). In North America, biodiversity assessments of temperate coniferous forest ecoregions range from *regionally outstanding* to *globally outstanding* [2]. Two extensive and important temperate coniferous forest ecoregions are the North Central Rockies Forests and South Central Rockies Forests, which include some of the world's most pristine wildlands, such as the Canadian Rockies, Northern Continental Divide Ecosystem, and the Greater Yellowstone Ecosystem. Although conifer diversity itself is low to moderate within these ecoregions, the composition and structure of each forest community type vary across steep moisture, topographic, and elevational gradients. Furthermore, wildfire is a recurring disturbance in Rocky Mountain forests, varying in intensity and return intervals across landscapes and leading to considerable community heterogeneity [3–6]. Variation among and within community types, as well as unique landscape community associations, are additive in contribution to regional biodiversity [7].

Community level biodiversity in these ecoregions, however, is not secure. These areas are epicenters for many of the forest health problems confronting western North American forests, including warming temperatures leading to reduced snowpack levels, earlier snowmelt [8], increased wildfire frequency [9], elevated tree mortality [10], as well as outbreaks of forest insects and pathogens [11,12], and ecosystem impacts of invasive pests, pathogens and plants [13]. These combined effects are predicted to push some forests over thresholds of resilience into new vegetation types, with a loss of forest ecosystem services [14].

Whitebark pine (*Pinus albicaulis* Engelm.) communities, which contribute significantly to western forest biodiversity, reach their distributional center of abundance in the North Central Rockies and South Central Rockies ecoregions [15]. They are impacted by a number of these forest health challenges across much of their range, but especially in the U.S. northwest and in the adjacent southwestern region of Canada [15,16]. In 2011, the species was designated a candidate under the U. S. Endangered Species Act and in 2012 declared endangered in Canada under the Species at Risk Act [17,18]. The two major factors cited in the species' decline were widespread mortality from white pine blister rust, a non-native disease caused by the fungal pathogen *Cronartium ribicola* J.C Fisch. in Rabh., which has spread nearly rangewide, and unprecedented outbreaks of mountain pine beetle (*Dendroctonus ponderosae* Hopkins). Other factors included altered fire regimes and advancing succession in seral communities from fire suppression efforts, and climate change [15,17,18].

Whitebark pine depends on the bird, Clark's nutcracker (*Nucifraga columbiana* (Wilson), for seed dispersal, which influences the pine's ecology, distribution, and population biology [19–21]. In late summer and fall, nutcrackers store ripe pine seeds throughout the montane landscape for future retrieval. Unretrieved seeds may germinate following snowmelt or rain, leading to regeneration (e.g., 22). Seed dispersal by nutcrackers to recently burned, disturbed, remote, or high elevation terrain, coupled with seedling tolerance of extreme conditions and harsh seedbeds, leads to whitebark pine regeneration early in succession and to whitebark pine establishment at the highest forest limits [20–22]. This coevolved, mutualistic seed dispersal system is related to the stone pine complex of Eurasia, which is dependent on seed dispersal by the Eurasian and large-spotted nutcrackers (*Nucifraga caryocates* L. and *N. multipunctata* Gould) [19].

Encompassing about 18° latitude and 21° longitude, whitebark pine has the largest latitudinal distribution of any North American five-needle white pine (Subgenus *Strobus*, Section *Quinquefoliae*, subsection *Strobus* [23]) but the narrowest elevational distribution comprising upper subalpine zone to treeline (Figure 1) [24]. The restricted elevational range results partly because whitebark pine is a "stress-tolerator" characterized by generally slow-growth, moderate shade-intolerance, and poor competitive abilities [24,25]. Whitebark pine community composition and dynamics change in relation to site characteristics: On comparatively warm, moist productive sites, communities are successional, with whitebark pine a minor component [26]. Although whitebark pine may persist into late seral stages in these communities, its overall basal area declines with time ([15] and references therein). These seral communities are disturbed periodically by fire. As sites become increasingly cold, and dry, with poorer soils, whitebark pine increases in community representation, becomes a self-replacing co-climax species, and then dominates as a self-replacing climax (late successional) species. Seral whitebark pine communities are most prevalent within the Rocky Mountain distribution, whereas the primary community types in the Pacific distribution are climax and treeline.

Figure 1. The distribution of whitebark pine (*Pinus albicaulis* Engelm.). (Source: Whitebark Pine Ecosystem Foundation [27]).

Across its range, whitebark pine functions as a foundation and keystone species [15,28]. A foundation species is regarded as " . . . disproportionately important to the continued maintenance of the existent community structure" ([29] p. 86). Loss of a foundational species profoundly alters structure, composition, and function [30]. Keystone species promote community diversity disproportionately in relation to their abundance through interactions with other species [31,32]. Foundation and keystone functions previously identified for whitebark pine are described in Table 1. Many of these functions result from highly effective seed dispersal by nutcrackers in combination with exceptionally hardy seedlings that rapidly develop deep tap roots and tolerate exposed, droughty sites [22,24,28]. Whitebark pine communities also provide ecosystem services, defined as ecosystem functions that benefit humans (e.g., [33]). Ecosystem services are divided into four general categories: provisioning services (e.g., food, timber), cultural services (e.g., recreation, inspiration), regulating services (e.g., preventing floods or disease), support services (e.g., clean water, pollination) [34]. Ecosystem services specific to whitebark pine are listed in Table 1.

Snowpack retention and snowmelt protraction are ecosystem services associated with whitebark pine subalpine forest and treeline communities (Table 1) [28,35]. Whitebark pine often grows in soils that hold little water and support sparse vegetation at elevations with high precipitation and snow accumulation, leading to copious run-off [35]. Because snow is the dominant form of precipitation in Rocky Mountain ecosystems, hydrologic processes are largely governed by snowpack dynamics, including snow deposition, redistribution, and runoff [41–44]. It has been inferred that high elevation whitebark pine communities create conditions that delay snowmelt and therefore downstream flow until late in the season, when it is most valuable for down-valley irrigation needs [28,35].

However, specific connections between whitebark pine communities and hydrologic processes lack empirical support.

Table 1. Keystone interactions (**K**), foundational functions (**F**), and ecosystem services provided by whitebark pine [15,16,28,35–38]. Ecosystem services to Native Americans [39,40].

Keystone Interactions and Foundational Functions
• Wide latitudinal distribution and environmental tolerances with a gradient of community types lead to forest, understory plant, and animal community diversity at local, regional, and range-wide scales. **K, F**
• Community structure provides wildlife shelter, perching and nesting sites, burrows, and food at high elevations. **K**
• Large, nutritious seeds provide an important wildlife food source for songbirds, upland game birds, squirrels and mice, bears, and foxes. **K**
• Hardiness of seedlings coupled with seed dispersal by Clark's nutcrackers leads to early whitebark pine regeneration after fire and other disturbance across slope aspects and elevations, which in turn mitigates harsh conditions and leads to community development through facilitation interactions. **F**
• Hardiness of seedlings leads to tolerance of harsh upper subalpine sites and establishment of whitebark pine. A less hardy tree may then establish in its lee, through facilitation, which fosters biodiversity. **K, F**
• Tolerance of harsh, windy sites, coupled with seed dispersal by Clark's nutcrackers to treeline microsites, leads to solitary whitebark pine facilitating leeward tree establishment, which enables tree island community development. **F**

Ecosystem services
• As the principal conifer growing at the highest elevations in the upper subalpine and at treeline, whitebark pine stands shade snowpack and slow snowmelt, which leads to protracted runoff and longer seasonal downstream flow. This potentially benefits agriculture and ranching.
• By growing at high elevations on harsh sites with rocky substrates, whitebark pine tree roots take up water, regulating flows, stabilize soils, and reduce soil erosion, potentially resulting in higher quality water with less sediment in downstream flows.
• Whitebark pine communities are associated with high elevation recreational activities in national forests, wilderness areas, national parks, and ski areas. On harsh sites, individual trees assume massive, twisted wind-sculpted forms, which to visitors symbolize survival against the elements and contribute uplifting aesthetics to the high-mountain experience.
• The large seeds of whitebark pine were an important traditional food for several Rocky Mountain Native American tribes. Tribes also used the inner bark of whitebark pine as a food source in spring or early summer.

In the North Central Rockies and South Central Rockies ecoregions, whitebark pine is a dominant krummholz species in many alpine-treeline ecotone (ATE) communities, especially on the harsh eastern front of the Rocky Mountains [38,45]. Within these communities, whitebark pine is not only the most abundant solitary species in these regions but also the species most often at the exposed windward edge of tree islands and thus the most common species to initiate tree island formation, a foundational function (Table 1). Whitebark pine is hardier under conditions of poor soils, wind, low snow cover, and drought than associated treeline conifers (e.g., [46]), and its leeward microsites provide a protective microclimate and reduced sky exposure, favoring the establishment of other trees [47].

The foundation and keystone functions of whitebark pine are generally recognized [15–18,28], as well is the range of forest biodiversity represented by whitebark pine subalpine communities [24,26,37]. We ask whether there may also be geographic variation in treeline community composition and structure, thus providing a more complete understanding of the range of forest types and community biodiversity represented by whitebark pine. Secondly, we review hydrologic processes at high elevations to determine whether ATE whitebark pine communities potentially retain snowpack and regulate local streamflows, which would provide inferential support for this ecosystem service in the absence of empirical evidence. Specifically, we examine both (1) geographic variation in the structure and composition of whitebark pine krummholz ATE communities and (2) geographic variation in dominant co-occurring "understory" plants, using information from our collective recent studies in the North Central and South Central Rockies. We also (3) review the literature for information on the contribution of treeline communities to snow retention and local hydrological processes, and connect this information to ATE whitebark pine communities. Furthermore, previous studies in the ATE across the distribution of whitebark pine indicate that *C. ribicola* is present in nearly every community examined [45], with the highest infection levels in the ATE of Glacier National Park [38,48,49]. Blister rust potentially threatens both the biodiversity supported by whitebark pine as a keystone and foundation species and the ecosystem services that whitebark pine communities provide. Finally, we explore the issue of whether restoration efforts are feasible or desirable for treeline communities.

2. Study Areas and Methods

Whitebark pine in the Rocky Mountains ranges from about 42° latitude in Wyoming, USA, to 54° latitude in British Columbia and Alberta, Canada. Summary data presented on the structure and composition of krummholz whitebark pine ATE communities come from ten studies either previously published or in preparation, spanning the latitudinal range of whitebark pine in the Rocky Mountains (Table 2, Figure 2).

Community assessment methods varied by study in plot size (225 m^2 *vs.* 500 m^2), number of plots (2–30), total area sampled (750 m^2 to 6750 m^2; median = 4500 m^2), and whether the plots were randomly or non-randomly distributed. Seven studies were based on a randomized distribution of study plots. For non-random sampling, plots were distributed within a defined aspect and slope at the highest elevations of the ATE that could be safely accessed; whereas, for random sampling, random points for plot initiation were selected *a priori* and stratified by aspect based on GIS or R [50]. We sampled non-randomly in the Willmore Wilderness Park, Stanley Glacier, and Gibbon Pass study areas, where surveys had to be completed at a specific location within a limited timeframe.

Sampled conifers were predominantly of krummholz growth form, or a mixture of krummholz with upright, flagged stems, and an occasional erect tree, especially alpine larch (*Larix lyallii* Parl.). Trees occurred either solitarily or in tree islands, where krummholz tree canopies were contiguous, and sometimes mixed with flagged branches and more upright forms. In all study areas, the following data were collected per plot: tree species richness S (not including shrub forms, such as *Juniperus* spp.), number of solitary trees per species, number of tree islands, species composition of each tree island by species presence (not by total number of each species), length of each tree island (to a maximum of 35 m), and species of the most windward tree (likely tree island initiator) for each tree island.

We calculated descriptive statistics for each study area using either Microsoft Excel or R [50] for plot level variables for numbers of solitary trees and numbers of tree islands per m^2, which we scaled up to densities per hectare, and tree island lengths. We report species richness among all solitary trees and tree islands and relative proportions of tree island initiators by species. The Shannon Index was used to calculate the community diversity of solitary trees for each study area:

$$H = -\sum_{i=1}^{S} (p_i) \, (log_2 \, p_i) \tag{1}$$

Information on understory plant dominants comes from our collaborative studies in these study areas. We reviewed the literature for information on the role of treeline conifer communities in snow retention, and thus provision of this important ecosystem service.

We tested data for normality. We performed linear regression to determine if the proportion of solitary trees represented by whitebark pine predicted the proportion of tree islands with whitebark pine as the initiator, and we performed Spearman rank correlation analysis to determine if tree island density correlated with tree island length [50].

Table 2. Study areas across the latitudinal range of whitebark pine in the Rocky Mountains, and sampling methods providing information on structure and composition of treeline communities.

Study Area	Latitude and Longitude	Elevation m	Aspect	Plot Size, N	Area Sampled; Distribution	Reference
Willmore Wilderness Park; Alberta, Canada	53°46′0″ N; 119°44′22″ W	1964–2175	SW, SE, E, NE	500 m²; N = 11	5500 m²; non-random	Tomback and Resler, in preparation [51]
Parker Ridge Banff National Park; Alberta, Canada	52°01′44″ N; 117°06′24″ W	2100	NE, NW, S, SE, SW	225 m²; N = 20	4500 m²; random	Resler et al. [52]; Resler et al., unpublished data [53]
Gibbon Pass; Banff National Park, Alberta, Canada	51°11′15″ N; 115°56′12″ W	2389–2430	SE, NE	500 m²; N = 3	1500 m²; non-random	Tomback et al. [45]
Stanley Glacier; Banff National Park, Alberta, Canada	51°11′13″ N; 116°02′51″ W	1969–1997	SW, NW	500 m²; N = 1; 250 m²; N = 1	750 m²; non-random	Tomback et al. [45]
Divide Mountain/White Calf Mountain; Glacier National Park, Blackfeet Indian Reservation, MT	48°39′25″ N; 113°23′45″ W; 48°38′20″ N; 113°24′08″ W	1920–2272	NE, NW, W, SW	225 m²; N = 30	6750 m²; random	Smith-McKenna et al. [48]
Line Creek; Custer National Forest, MT	45°01′47″ N; 109°24′09″ W	2950	NE	225 m²; N = 30	6750 m²; random	Smith-McKenna et al. [48]
Tibbs Butte; Shoshone National Forest, WY	44°56′28″ N; 109°26′39.69″ W	2983–3238	E, NE	225 m²; N = 12	2700 m²; random	Wagner et al., in preparation [54]
Paintbrush Divide/Holly Lake; Grand Teton National Park, WY	43°47′34″ N; 110°47′54″ W	3055–3289	NW, SW, NE	225 m²; N = 20	4500 m²; random	Resler and Shao, unpublished data [55]
Hurricane Pass/Avalanche Basin; Grand Teton National Park, WY	43°43′41″ N; 110°51′02″ W	3045–3204	NW, SW, NE	225 m²; N = 20	4500 m²; random	Resler and Shao, unpublished data [55]
Christina Lake; Shoshone National Forest, WY	42°35′35″ N; 108°58′24″ W	3200–3400	SW, SE, NE	225 m²; N = 30	6,750 m²; random	Wagner et al., in preparation [54]

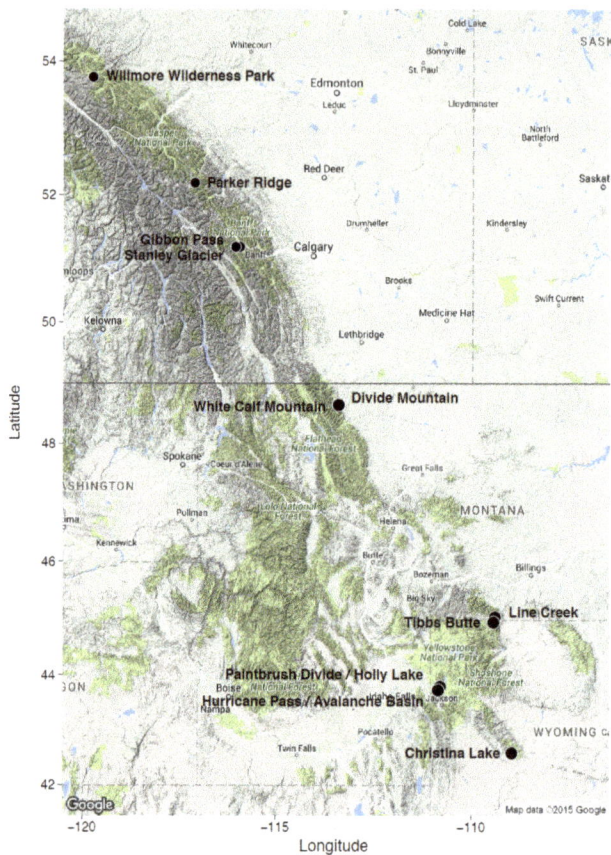

Figure 2. The locations of the ten alpine-treeline ecotone (ATE) community study areas encompassing the geographic range of whitebark pine in the Northern and Central Rocky Mountains. These areas provided the comparative data used in this study. See Table 2 for additional information about each study area.

3. Results

3.1. Geographic Variation in ATE Community Composition and Structure

The ATE communities in all study areas comprised both small, scattered solitary krummholz trees interspersed among tree islands and predominantly krummholz tree islands. Solitary trees occurred in open terrain among tree islands but also as non-contiguous satellites around tree islands. Median densities of solitary trees per hectare, with all species combined, ranged from 160.00 trees/ha at Gibbon Pass to 977.78 trees/ha at Divide Mountain (Table 3). With the exception of Gibbon Pass, the highest densities of solitary trees occurred within the northern study areas. The median densities of tree islands ranged from 0.00/ha at both Tibbs Butte and Christina Lake to 130.00/ha at Stanley Glacier, but half of the median values were *ca.* 80/ha (Table 3). Again, the highest median densities were generally in the northern study areas. At Stanley Glacier, Divide Mountain, Paintbrush Divide/Holly Lake, and Hurricane Pass/Avalanche Basin, we encountered one or more tree islands longer than 35 m in length (Table 3). Median tree island lengths ranged from 2.59 m to 7.68 m among the study areas, with no obvious trends.

Table 3. Tree species richness, conifer composition, solitary tree density and tree island density, tree island length, and understory dominants for the ten study areas listed in Table 2. Conifer abbreviations are as follows: PIAL = whitebark pine (*Pinus albicaulis*), ABLA = subalpine fir (*Abies lasiocarpa*), PIEN = Engelmann spruce (*Picea engelmannii*), PICO = lodgepole pine (*Pinus contorta*), LALY = alpine larch (*Larix lyallii*), and PSME = Douglas-fir (*Pseudotsuga menziesii*).

Location	Tree Species Richness S Solitary	Tree Species Richness S Tree Islands	Median, Min, Max, No. Solitary Trees/ha, total Solitary Trees n	Shannon Diversity Index for Solitary Trees	Median, Min, Max, Tree Islands/ha, Total Tree Islands n	Median, Min, Max Length m, Total Tree Islands n	Dominant Understory Species
Willmore Wilderness Park	4 PIAL ABLA PIEN PICO	4 PIAL ABLA PIEN PICO	780.00 140.00 2080.00 n = 485	0.724	80.00 20.00 200.00 n = 49	2.59 0.10 9.60 n = 48	*Dryas integrifolia* Vahl *Cassiope mertensiana* (Bong.) G. Don *Empetrum nigrum* L. *Phyllodoce glanduliflora* (Hook.) Coville *Rhododendron albiflorum* Hook. *Menziesia ferruginea* Sm. *Vaccinium membranaceum* Douglas ex Torr.
Parker Ridge	3 PIAL ABLA PIEN	3 PIAL ABLA PIEN	777.78 88.89 2400.00 n = 321	0.830	88.89 0 222.22 n = 49	2.70 0.04 19.90 n = 49	*Dryas octopetala* L. *Silene acaulis* (L.) Jacq.
Gibbon Pass	4 PIAL ABLA PIEN LALY	4 PIAL ABLA PIEN LALY	160.00 120.00 1760.00 n = 103	1.312	80.00 80.00 100.00 n = 12	4.35 0.65 11.60 n = 12	*Dryas octopetala* L. *Silene acaulis* (L.) Jacq.
Stanley Glacier	3 PIAL ABLA PIEN	3 PIAL ABLA PIEN	720.00 240.00 1200.00 n = 32	0.769	130.00 60.00 200.00 n = 8	7.34* 1.00 >35.00 n = 8	*Dryas octopetala* L. *Silene acaulis* (L.) Jacq.
Divide Mountain	4 PIAL ABLA PIEN PSME	4 PIAL ABLA PIEN PSME	977.78 88.89 3911.11 n = 367	0.778	111.00 0.00 355.55 n = 84	5.40 0.40 >35.00 n = 84	*Dryas octopetala* L. *Arctostaphylos uva-ursi* (L.) Spreng. *Hedysarum sulphurescens* Rydb. *Oxytropis sericea* Nutt. *Silene acaulis* (L.) Jacq.
White Calf Mountain							*Arctostaphylos uva-ursi* (L.) Spreng. *Achillea millefolium* L. *Hedysarum sulphurescens* Rydb. *Potentilla diversifolia* Lehm.

Table 3. Cont.

Location	Tree Species Richness S		Median, Min, Max, No. Solitary Trees/ha, total Solitary Trees n	Shannon Diversity Index for Solitary Trees	Median, Min, Max, Tree Islands/ha, Total Tree Islands n	Median, Min, Max Length m, Total Tree Islands n	Dominant Understory Species
	Solitary	Tree Islands					
Line Creek	5 PIAL ABLA PIEN PSME PICO	5 PIAL ABLA PIEN PSME PICO	533.33 88.89 1733.33 n = 244	0.695	88.89 0 444.44 n = 62	4.50 0.76 16.88 n = 62	Ageseris glauca (Pursh) Raf. Aster alpigenus (Torr. & A. Gray) A. Gray Geum rossii (R. Br.) Ser. Potentilla diversifolia Lehm. Lupinus argenteus Pursh
Tibbs Butte	2 PIAL PIEN	3 PIAL ABLA PIEN	222.22 44.44 1422.22 n = 88	0.249	0.00 0.00 88.89 n = 3	7.68 0.76 14.07 n = 3	Geum rossii (R. Br.) Ser. Potentilla diversifolia Lehm. Saxifrage spp. Graminoids
Paintbrush Divide/Holly Lake	3 PIAL ABLA PIEN	3 PIAL ABLA PIEN	550.71 0.00 1497.93 n = 282	0.828	88.11 0.00 308.40 n = 45	3.35 0.11 >35.00 n = 45	Arctostaphylos uva-ursi (L.) Spreng. Myosotis asiatica (Vesterg.) Schischkin & Sergievskaja Senecio spp. Silene acaulis (L.) Jacq.
Hurricane Pass/Avalanche Basin	4 PIAL ABLA PIEN PSME	4 PIAL ABLA PIEN PSME	242.31 0.00 793.02 n = 138	1.069	66.08 0.00 352.45 n = 45	7.30 0.98 >35.00 n = 45	Dryas octopetala L. Senecio spp. Silene acaulis (L.) Jacq. Graminoids
Christina Lake	3 PIAL ABLA PIEN	3 PIAL ABLA PIEN	244.44 0.00 2000.00 n = 247	0.421	0.00 0.00 444.44 n = 41	3.39 1.14 29.07 n = 41	Geum rossii (R. Br.) Ser. Pteryxia hendersonii (J.M. Coult. & Rose) Mathias & Constance Phlox pulvinata (Wherry) Cronquist Silene acaulis (L.) Jacq.

Among study areas, conifer species richness S for solitary trees ranged from 2 to 5 and for tree islands from 3 to 5, with the highest richness in the Line Creek study area (Table 3). With the exception of the Gibbon Pass and Tibbs Butte study areas, solitary trees and tree islands were composed primarily of whitebark pine (PIAL), subalpine fir (*Abies lasiocarpa* (Hook.) Nutt., ABLA), and Engelmann spruce (*Picea engelmannii* Parry ex Engelmann, PIEN). Small numbers of other species occurred in several study areas: lodgepole pine (*Pinus contorta* Douglas ex Loudon, PICO) in Willmore Wilderness Park and Line Creek, alpine larch (*Larix lyallii* Parl., LALY) in Gibbon Pass, and Douglas-fir (*Pseudotsuga menziesii* (Mirb.) Franco, PSME) in Divide Mountain and Line Creek (Table 3).

Among solitary trees, whitebark pine was the majority species in nine of the ten study areas, ranging from 93.2% at Tibbs Butte to 23.3% at Gibbon Pass—the latter an outlier in community composition (Figure 3). Species comprising the other solitary trees varied in relative abundance (Figure 3). At Gibbon Pass, Engelmann spruce dominated solitary trees, accounting for 37.9%. In other study areas, Engelmann spruce ranged in percent occurrence among solitary trees from 5.3% at Christina Lake to 34.4% at Stanley Glacier. Subalpine fir ranged from 5.7% at Christina Lake to 23.2% at Hurricane Pass/Avalanche Basin. At Gibbon Pass, alpine larch was the second most prevalent solitary tree after spruce, and at Line Creek, lodgepole pine was more common than subalpine fir. The Gibbon Pass and Hurricane Pass/Avalanche Basin study areas had the highest Shannon diversity indices for the solitary tree component of the ATE community, primarily because of greater evenness in relative occurrence among conifers (Table 3). The Tibbs Butte study area had the lowest Shannon diversity index, because of dominance by whitebark pine and a low species richness of 2. One or more whitebark pine trees occurred across the highest proportion of tree islands than any other conifer species in all study areas except Parker Ridge and Gibbon Pass (Figure 4).

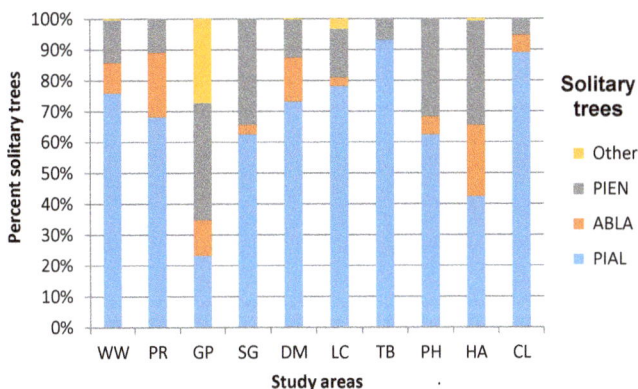

Figure 3. Percent composition of the solitary conifer tree community for each study area. See Table 3 for conifer species listed under "Other" for each study area. Conifer abbreviations are as follows: PIAL = whitebark pine (*Pinus albicaulis*), ABLA = subalpine fir (*Abies lasiocarpa*), PIEN = Engelmann spruce (*Picea engelmannii*), PICO = lodgepole pine (*Pinus contorta*), LALY = alpine larch (*Larix lyallii*), and PSME = Douglas-fir (*Pseudotsuga menziesii*). Study areas are abbreviation as follows: WW = Willmore Wilderness, PR = Parker Ridge, GP = Gibbon Pass, SG = Stanley Glacier, DM = Divide Mountain/White Calf Mountain, LC = Line Creek, TB = Tibbs Butte, PH = Paintbrush Divide/Holly Lake, HA = Hurricane Pass/Avalanche Basin, and CL = Christina Lake.

The relative proportions of tree islands with subalpine fir and Engelmann spruce varied across study areas from Line Creek north, but were comparable in study areas from Tibbs Butte south. Median tree island length varied from 2.59 m in the Willmore Wilderness study area to 7.68 m at Tibbs Butte; no latitudinal gradient was apparent, and median density of tree islands did not predict length of tree islands ($\rho = 0.191$ (95% CI: $-0.572, 0.881$), $N = 10$, $p = 0.598$).

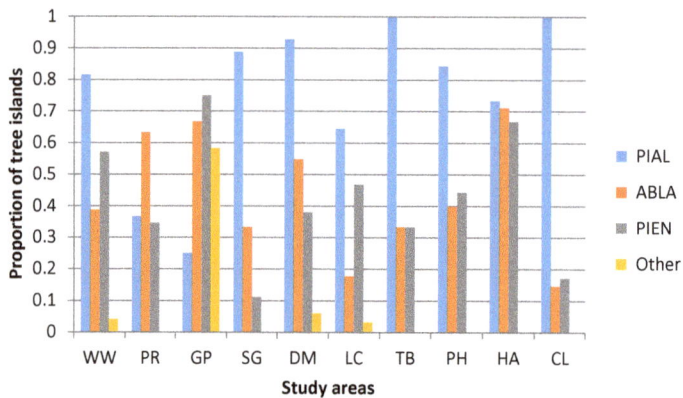

Figure 4. The proportion of tree islands within a study area with one or more individuals of each species. See Table 3 for conifer species listed under "Other", and Figure 3 for conifer species abbreviations and study area abbreviations.

Whitebark pine was the most frequently occurring windward conifer in tree islands, and thus presumably the majority tree island initiator, within five study areas: Stanley Glacier, Divide Mountain, Line Creek, Tibbs Butte, and Christina Lake (Figure 5). We examined whether the relative proportion of whitebark pine as a solitary tree predicted its relative proportion as a tree island initiator across study areas. We found that proportional abundance of whitebark pine among solitary trees predicted its proportional abundance as a tree island initiator ($F = 8.724$, $r = 0.722$, $R^2 = 0.522$, $df = 8$, $p = 0.018$). Several study areas, such as Willmore Wilderness Park, Parker Ridge, and Paintbrush Divide/Holly Lake, had high proportions of whitebark pine among solitary trees but not among tree island initiators (Figure 3). We tested the same data sets for subalpine fir and Engelmann spruce and found that proportional representation of these conifers as solitary trees did not as strongly predict their proportional abundance as tree island initiators ($F = 5.192$, $r = 0.627$, $R^2 = 0.3936$, $df = 8$, $p = 0.0522$, and $F = 2.645$, $r = 0.498$, $R^2 = 0.2485$, $df = 8$, $p = 0.143$, respectively).

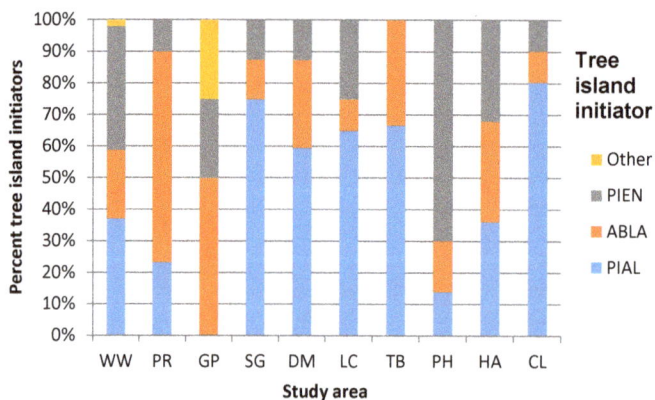

Figure 5. Relative proportional abundance of whitebark pine as a tree island initiator, as inferred by its leading windward position in tree islands, within each study area. See Table 3 for conifer species listed under "Other", and Figure 3 for conifer species abbreviations and study area abbreviations.

3.2. Geographic Variation in ATE Understory Community Composition

The ATE understory vascular plant community varied among and within regions in composition (Table 3). The Willmore Wilderness Park study area understory community included lichens, especially *Cladonia* spp. and *Stereocaulon* spp. (J. Gould, personal communication), as well as large patches of multi-species heath communities (*Cassiope mertensiana, Empetrum nigrum, Phyllodoce glanduliflora*), which indicate substantial moisture from a maritime climatic influence [26,56]. Widely-distributed species shared among several study areas included *Dryas octopetala, Arctostaphylos uva-ursi*, and *Silene acaulis*; and, all study areas featured low-growing *Salix* spp., not listed in Table 3. The substrate at the Stanley Glacier and Gibbon Pass study areas comprised loose, unstable shale fragments from weathering, possibly contributing to the sparse and patchy understory at these sites.

Many of our treeline study areas are located east of the Continental Divide and exposed to the extremes of a continental climate. Exceptions include the Stanley Glacier study area, which lies west of the Continental Divide; the Divide Mountain/White Calf study area, which experiences Pacific air masses and precipitation levels similar to those west of the Divide [57]; and, the Paintbrush Divide/Holly Lake and Hurricane Pass/Avalanche Basin study areas, which occur west of the Divide within the Teton Range, a region which receives relatively heavy snowfall [56]. The latter two southern study areas included *Arctostaphylos uva-ursi* and *Dryas octopetala* in the understory community—two species characteristic of the more mesic treelines of the northern regions. The annual snowfall in the southern and eastern Wind River Range, the location of the Christina Lake study area, is half that of the Teton Range, and the region is characterized as semi-arid [56].

3.3. Snow Retention As an Ecosystem Service

We conducted a literature search on snowpack deposition, redistribution, and snowmelt runoff to determine whether treeline communities in general influence hydrologic processes, and whether we can extrapolate these processes to ATE whitebark pine communities. Our search revealed a long history of literature describing the spatial patterns of snow redistribution and accumulation (Table 4). These studies, which originated with the implementation of snow fences to make winter highway driving safer (e.g., [58,59]), led to the development of algorithms describing snow redistribution by wind, which indicated that objects act as snow-traps, effectively forcing snow distribution in their lee [58,60–64]. Watershed studies in arid, western plant communities indicated that strategically-placed snow fences increased snow accumulation, snowmelt discharge and the length of the flow period ([65] and references therein). Subsequent subalpine and treeline forest observational studies verified that snow depth is positively associated with tree distribution, as observed for snow fences [43,66–70] (Figure 6, Table 4). Snow depth influences other hydrological processes, including snow water equivalent (SWE), runoff initiation date, snow cover season length, and the spatial pattern of snow disappearance[41,69–72] (Table 4). Delayed runoff initiation prolongs snowdrift disappearance in areas with deeper snow [70].

Snow accumulation and distribution drive other ecosystem processes at treeline including allochthonous deposition of nutrients and herbaceous community structure [68,73–78] (Table 4). Treeline environments experience similar rates of inorganic nutrient deposition rates as subalpine forest, and higher rates than treeless alpine tundra [68]. Snowpack depth and snow cover duration also influence plant community composition and spatial distribution at both local and landscape scales [73–76], because snow-free date determines growing season length, water availability, and soil temperatures [72,74].

Table 4. References describing snowpack deposition, redistribution, snowmelt run-off, and the role of high elevation tree communities in these processes. The processes are summarized with respect to potential snow retention ecosystem services provided by whitebark pine communities in the alpine-treeline ecotone.

Topic	References
Algorithms and highway safety research indicate wind distributes snow leeward of objects.	[58,60–62]
Field measurements demonstrate increased snow accumulation, snow depth, and snow water equivalent leeward of trees.	[42,66–69,80]
Increased snow depth positively correlates with snow water equivalent (SWE).	[69,70]
Increased snow depth is associated with delayed runoff initiation and longer snow persistence.	[41,70–72]
Snow depth and snowpack persistence predict herbaceous vegetation community distribution.	[73–76]
Snow accumulation is positively associated with soil nutrient deposition.	[68,77,78]
Field based assessment of a whitebark pine ecosystem indicates vegetation influences snow accumulation and spatial distribution.	[66]
Summary of processes	
1. Objects, such as trees and rocks, act as snowtraps, forcing snow redistribution in their lee.	
2. The distribution of trees on the landscape positively influences snow depth.	
3. An increase in snow depth directly increases downstream water yield.	
4. Whitebark pine is abundant at treeline and initiates tree islands, creating snow traps, and thus influences local hydrological processes.	

Thus, the presence of tree stands and tree islands influences local hydrologic processes, leading to deeper snow deposition in their lee and directing the spatial distribution of snow (Figure 6). Because whitebark pine is the dominant conifer in many ATE communities throughout the Rocky Mountains and often initiates tree island development [38,45,79], whitebark pine communities influence snow depth in their lee and contribute to the spatial heterogeneity of herbaceous community composition at treeline. Even if whitebark pine is not locally important in tree island initiation, it is typically an important structural component of tree islands.

(a)

Figure 6. *Cont.*

(**b**)

Figure 6. (**a**) Ribbon forest ATE community above Christina Lake, WY, with whitebark pine a major component after snowmelt completion in the surrounding treeless matrix (Photo: E.R. Pansing); (**b**) Whitebark pine ATE community from Divide Mountain, MT, with snowdrift in lee of tree island stringer (Photo: D.F. Tomback).

3.4. Blister Rust Incidence Regionally and Within Treeline Communities

White pine blister rust now occurs throughout whitebark pine's Rocky Mountain distribution in subalpine forest communities, with lower incidence at higher latitudes and east of the Continental Divide (Table 5) ([81,82], see Table 1.1). Repeated surveys on permanent plots indicate widespread infection in both the Central Rocky Mountains and Northern Rocky Mountains and tree mortality from the disease [83,84]. White pine blister rust is present across all regions at treeline but at low incidence at the northern and southern distributional limits, where climatic conditions are less favorable for the disease (Table 5) [85,86].

Table 5. Incidence of infection by white pine blister rust (BR) in subalpine forests in the Central and Northern Rocky Mountains, and incidence of blister rust at treeline across the Rocky Mountains.

Location	Overall Infection Incidence; Mortality Notes	Reference
Subalpine: Greater Yellowstone Area	20%–30% mortality indicated	[83]
Subalpine: Canadian Rocky Mountains	52% 28% of trees, primarily BR	[81]
Treeline: east slope Glacier National Park: Divide Mountain and Lee Ridge	33.7% multiple symptoms; 24.3% confirmed cankered; mortality in both study areas: 6 dead trees on 3 transects ; 6 dead trees on 4 transects; confirmed BR	[38]
Treeline: Willmore Wilderness Park	1.1% (confirmed cankers)	Tomback and Resler, in preparation [51]
Treeline: Glacier National park, 6 study areas	47% (0%–100% per plot)	[48]
Treeline: Divide Mountain; Line Creek	23.6% (confirmed cankers); 19.2% (confirmed cankers)	[49]
Treeline: Gibbon Pass; Stanley Glacier; Tibbs Butte	0% 16.2% (multiple symptoms); 10.8% (confirmed cankers); 0% (on transects; presence off transects)	[45]
Treeline: Paintbrush Divide/Holly Lake; Hurricane Pass/Avalanche Basin	17.2%; 14.9%	Resler and Shao unpublished data [55]
Treeline: Tibbs Butte; Christina Lake	<10% blister rust in both areas	Wagner *et al.*, in preparation [54]

The highest incidences of blister rust in subalpine whitebark pine communities occur in the Northern Continental Divide Ecosystem (e.g., Glacier and Waterton Lakes National Parks) [81], where high treeline infection levels also reflect these trends (Table 5). In the Greater Yellowstone Area, infection at the Line Creek study area was comparable to levels generally within subalpine communities (Table 5) [83]. The outlook for blister rust in North America is continued spread and regional intensification throughout the collective ranges of whitebark pine and other five-needle white pines [15,86].

4. Discussion

4.1. Potential Impacts to ATE Whitebark Pine Communities from White Pine Blister Rust

Tomback and Resler [79] indicated that *Cronartium ribicola* could disrupt ATE communities in two ways. First, loss of seed production in subalpine whitebark pine communities from tree mortality and damage to tree canopies reduces the likelihood of seed dispersal from the subalpine communities by Clark's nutcrackers to treeline [20,87–89]. Conifers in the ATE only rarely produce seed cones, and often the seeds are not viable; the seed sources for ATE conifer regeneration are trees from lower elevations [90]. Secondly, blister rust infection at treeline damages and kills trees, potentially reducing the availability of whitebark pine to serve as a tree island initiator, which is especially important on the harshest and highest treeline sites, but also potentially eliminates whitebark pine as a frequent component of tree islands [38]. In this study, we found a direct relationship between the proportion of solitary whitebark pine and the proportion of tree islands initiated by whitebark pine, suggesting that a decline in whitebark pine could also lead to altered community composition and reduction in tree island density [79]. Understory plant composition may change as well, since patterns of snow retention and nutrient deposition may be impacted. Reduction in tree island size or density may reduce the effectiveness of snow retention in the ATE, which will further influence community composition.

Given the functional role of whitebark pine as a tree island initiator, mortality from blister rust may alter treeline response to climate change [79]. If loss of seed dispersal rrestricts the upward elevational movement of whitebark pine in response to increasing temperatures, or whitebark pine established above the limits of the current ATE succumbs to blister rust, then opportunities for tree island development may be reduced. This could lead to the perception that treeline is either slow to respond to warming trends or is not moving upwards in response to warming [79]. Agent-based modeling of the scenario "climate warming and disease" indicated that over a 500 to 800 year timeframe, whitebark pine in the presence of blister rust would decline by over 60%, Engelmann spruce exceeded whitebark pine in abundance after year 630, but the area occupied by tundra increased as the area occupied by conifers declined [91].

4.2. Geographic Variation in ATE Whitebark Pine Communities

Although ATE whitebark pine communities are principally composed of three conifer species throughout their Rocky Mountain range, we see a difference in structure and composition among study areas within regions and across regions, primarily in response to topographic and climatic variation, but also in response to biotic factors, such as seed availability and site suitability [90]. Median density of solitary trees varies more than four-fold across study areas, and the median density of tree islands varies from 0 to 130 tree islands/ha. In different study areas, additional conifer species are part of the ATE community, and the relative species composition of both solitary trees and tree islands varies among them. Furthermore, the understory communities may share some of the same plant species, but the associations and even relative abundances vary across study areas. Thus, whitebark pine treeline communities are highly variable, and based on working definitions of biodiversity [7,92], even highly biodiverse.

4.3. The role of Whitebark Pine Communities In Snow Retention

Trees within the ATE help determine snow spatial distribution across the landscape by increasing snow depth in their lee [42,62,67,68,80]. Snow depth affects snowpack characteristics, including SWE, timing of snowmelt initiation and runoff, and snow disappearance dates [42,58,62,70,73,80]. Snow distribution also influences ecosystem structure through nutrient deposition and herbaceous community composition and distribution by prolonging snowmelt initiation and snowmelt dates [68,77].

Because the presence of trees at high elevations impacts hydrologic processes, whitebark pine and whitebark pine communities potentially serve important roles in increasing snow depth locally and determining the spatial distribution of snow. Whitebark pine's tolerance of windswept, droughty sites [24,26]) combined with its role as a tree island initiator and major structural component of tree islands suggest that, in some watersheds in particular, whitebark pine not only assumes an important role in mediating hydrological processes but also influences local ecosystem processes beyond runoff dynamics, including nutrient capture and distribution, vegetation community composition, and vegetation spatial distribution (e.g., [65,68,73–78]). Although these hydrological processes are a logical extrapolation of known mechanisms, the role of treeline whitebark pine communities should be substantiated by empirical data.

4.4. Management Implications

Forest managers in the U.S. and Canada recognize that the growing losses of whitebark pine impact forest community biodiversity and lead to declining ecosystem function and services [17,18]. Restoration strategies for subalpine whitebark pine communities now exist or are under development [82,93,94]. For example, Keane *et al.* [82] presented general approaches for restoring whitebark pine, and Keane *et al.* [94] developed recommendations to mitigate climate change impacts.

We offer ideas for a two-pronged restoration strategy for ATE whitebark pine communities, which are experiencing mortality from blister rust [38]. First, existing restoration protocols apply to whitebark pine communities at the upper limits of the subalpine zone, which are the seed sources for ATE communities [87,90]. The most effective restoration treatment is to plant seedlings grown from seed sources previously identified as having some degree of genetic blister rust-resistance in areas with high tree mortality, such as after recent fires, after pine beetle outbreaks, or with high rates of blister rust [82]. This potentially increases the representation of resistance in populations, but is difficult to implement at large scales. Treatments such as thinning and prescribed burning reduce competition from faster-growing shade-tolerant trees [95]; these treatments are generally not appropriate for whitebark pine stands just below the ATE, which tend to be open, self-replacing climax communities.

Secondly, whitebark pine seedlings should be planted throughout the ATE where blister rust incidence is high to compensate for solitary krummholz tree mortality and reduced rates of seed dispersal from the upper subalpine communities. Although whitebark pine may grow without a protective microsite or nurse object in the ATE [46], survival is more likely if whitebark pine is planted leeward of a protective nurse object, such as a rock or shrub. A logistically efficient alternative to planting may be sowing seeds with potential genetic resistance to blister rust in protected sites in the ATE. Pansing *et al.* [96] simulated nutcracker seed dispersal by sowing caches of whitebark pine seeds in the ATE at the Tibbs Butte and White Calf Mountain study areas. Although 60% of the sown caches in both study areas were pilfered by rodents, the remaining seeds experienced greater than 60% germination in both study areas and one-year seedling survival of about 70% at Tibbs Butte and 30% survival at White Calf. Seedlings near rocks and trees had higher survival rates than in unprotected sites. These survival rates are sufficiently reasonable to suggest that for the ATE, direct seed sowing may be more cost effective and logistically feasible than growing seedlings in nurseries for two or three years and contracting planting crews, which is the current protocol.

5. Conclusions

Treeline whitebark pine communities across the Rocky Mountain distribution of whitebark are highly variable, and thus highly biodiverse, according to an inclusive definition of biodiversity [7]. Treeline communities perform the important ecosystem service of snow retention and protraction of snowmelt. Whitebark pine is a major structural component of many Rocky Mountain treeline communities, and in a subset of communities is often the principal initiator of tree islands. Loss of whitebark pine to white pine blister rust will alter treeline community composition and structure, and potentially alters response to climate warming. In addition, reduction in density of tree islands on harsh sites may reduce snow deposition and retention in these communities, altering local hydrology. We suggest that current whitebark pine restoration strategies should include planting blister rust-resistant whitebark pine seedlings at the highest subalpine elevations but also sowing seeds within the ATE at suitable sites.

Acknowledgments: We thank Joyce Gould, Alberta Parks, for information on treeline understory plant communities in Willmore Wilderness Park. The studies that provided data for this paper were funded by the National Science Foundation (grant 808548 to LMR, DFT, and G. P. Malanson); University of Colorado Denver, Faculty Development Grant (to DFT); Shoshone National Forest (grant FS-1500-17B, to DFT); and, University of Wyoming and U.S. Department of Interior, National Park Service (award ID 1002614E-VATECH to LMR). Manuscript preparation in part was supported by a Charles Bullard Harvard Forest Fellowship, Harvard University, to DFT.

Author Contributions: Diana F. Tomback, Lynn M. Resler, Robert E. Keane, and Elizabeth R. Pansing designed the study. Lynn M. Resler and Aaron C. Wagner provided unpublished data for use in the analysis; Diana F. Tomback and Andrew J. Andrade compiled data; Aaron C. Wagner and Andrew J. Andrade performed data analyses. Diana F. Tomback, Lynn M. Resler, Elizabeth R. Pansing, Robert E. Keane, and Andrew J. Andrade wrote the manuscript.

Conflicts of Interest: The authors declare no conflict of interest.

References

1. Pan, Y.; Birdsey, R.A.; Phillips, O.L.; Jackson, R.B. The structure, distribution, and biomass of the world's forests. *Annu. Rev. Ecol. Syst.* **2013**, *44*, 593–622. [CrossRef]
2. Ricketts, T.H.; Dinerstein, E.; Olson, D.M.; Loucks, C.J.; Eichbaum, W.; DellaSala, D.; Kavanagh, K.; Hedao, P.; Hurley, P.T.; Carney, K.M.; *et al. Terrestrial Ecoregions of North America: A Conservation Assessment*; Island Press: Washington, WA, USA; Covelo, CA, USA, 1999.
3. Habeck, J.R.; Mutch, R.W. Fire-dependent forests in the northern Rocky Mountains. *Quat. Res.* **1973**, *3*, 408–424. [CrossRef]
4. Romme, W.H.; Despain, D.G. The long history of fire in the Greater Yellowstone Ecosystem. *West. Wildlands* **1989**, *15*, 10–17.
5. Fischer, W.C.; Clayton, B.D. *Fire Ecology of Montana Forest Habitat types East of the Continental Divide*; General Technical Report INT-141; USDA Forest Service Intermountain Forest and Range Experiment Station: Ogden, UT, USA, 1983.
6. Turner, M.G.; Hargrove, W.W.; Gardner, R.H.; Romme, W.H. Effects of fire on landscape heterogeneity in Yellowstone National Park, Wyoming. *J. Veg. Sci.* **1994**, *5*, 731–742. [CrossRef]
7. Noss, R.F. Indicators for monitoring biodiversity: A hierarchical approach. *Conserv. Biol.* **1990**, *4*, 355–364. [CrossRef]
8. Pederson, G.T.; Graumlich, L.J.; Fagre, D.B.; Kipfer, T.; Muhlfeld, C.C. A century of climate and ecosystem change in Western Montana: What do temperature trends portend? *Clim. Chang.* **2011**, *98*, 133–154. [CrossRef]
9. Westerling, A.L.; Hidalgo, H.G.; Cayan, D.R.; Swetnam, T.W. Warming and earlier spring increase western U.S. forest wildfire activity. *Science* **2006**, *313*, 940–943. [CrossRef] [PubMed]
10. Van Mantgem, P.J.; Stephenson, N.L.; Byrne, J.C.; Daniels, L.D.; Franklin, J.F.; Fulé, P.Z.; Harmon, M.E.; Larson, A.J.; Smith, J.M.; Taylor, A.H.; *et al.* Widespread increase of tree mortality rates in the western United States. *Science* **2009**, *323*, 521–524. [CrossRef] [PubMed]

11. Raffa, K.F.; Aukema, B.H.; Bentz, B.J.; Carroll, A.L.; Hicke, J.A.; Turner, M.G.; Romme, W.H. Cross-scale drivers of natural disturbances prone to anthropogenic amplification: The dynamics of bark beetle eruptions. *BioScience* **2008**, *58*, 501–517. [CrossRef]

12. Weed, A.S.; Ayres, M.P.; Hicke, J.A. Consequences of climate change for biotic disturbances in North American forests. *Ecol. Monogr.* **2013**, *83*, 441–470. [CrossRef]

13. Roy, B.A.; Alexander, H.M.; Davidson, J.; Campbell, F.T.; Burdon, J.J.; Sniezko, R.; Brasier, C. Increasing forest loss worldwide from invasive pests requires new trade regulations. *Front. Ecol. Environ.* **2014**, *12*, 457–465. [CrossRef]

14. Millar, C.I.; Stephenson, N.L. Temperate forest health in an era of emerging megadisturbance. *Science* **2015**, *349*, 823–826. [CrossRef] [PubMed]

15. Tomback, D.F.; Achuff, P. Blister rust and western forest biodiversity: Ecology, values and outlook for white pines. *For. Pathol.* **2010**, *40*, 186–225. [CrossRef]

16. Tomback, D.F.; Achuff, P.; Schoettle, A.W.; Schwandt, J.; Mastrogiuseppe, R.J. The Magnificent High-Elevation Five-Needle White Pines: Ecological Roles and Future Outlook. In *Proceedings High-Five Symposium: The Future of High-Elevation Five-Needle White Pines in Western North America*; Keane, R.E., Tomback, D.F., Murray, M.P., Smith, C.M., Eds.; USDA Forest Service, Rocky Mountain Research Station: Fort Collins, CO, USA, 2011; pp. 2–28.

17. U.S. Fish and Wildlife Service. Endangered and threatened wildlife and plants; 12-month finding on a petition to list *Pinus albicaulis* as Endangered or Threatened with critical habitat. *Fed. Regist.* **2011**, *76*, 42631–42654.

18. Government of Canada. Order amending Schedule 1 to the Species at Risk Act. *Canada Gazette*, 2012; Part II. Vol. 146. No. 14, SOR/2012–113. Available online: http://www.sararegistry.gc.ca/virtual_sara/files/orders/g2-14614i_e.pdf (accessed on 20 June 2012).

19. Tomback, D.F.; Linhart, Y.B. The evolution of bird-dispersed pines. *Evolut. Ecol.* **1990**, *4*, 185–219. [CrossRef]

20. Tomback, D.F. Clark's Nutcracker: Agent of regeneration. In *Whitebark Pine Communities: Ecology and Restoration*; Tomback, D.F., Arno, S.F., Keane, R.E., Eds.; Island Press: Washington, DC, USA, 2001; pp. 89–104.

21. Tomback, D.F. The impact of seed dispersal by Clark's nutcracker on whitebark pine: Multi-Scale perspective on a high mountain mutualism. In *Mountain Ecosystems: Studies in Treeline Ecology*; Broll, G., Keplin, B., Eds.; Springer: Berlin, Germany, 2005; pp. 181–201.

22. Tomback, D.F.; Anderies, A.J.; Carsey, K.S.; Powell, M.L.; Mellmann-Brown, S. Delayed seed germination in whitebark pine and regeneration patterns following the Yellowstone fires. *Ecology* **2001**, *82*, 2587–2600. [CrossRef]

23. Gernandt, D.S.; Geada López, G.G.; Ortiz Garcia, S.; Liston, A. Phylogeny and classification of *Pinus. Taxon* **2005**, *54*, 29–42. [CrossRef]

24. Arno, S.F.; Hoff, R.J. *Pinus albicaulis* Engelm. Whitebark pine. In *Silvics of North America; Conifers*; Burns, R.P., Honkala, B.H., Eds.; USDA Forest Service: Washington, DC, USA, 1990; Volume 1, pp. 268–279.

25. McCune, B. Ecological diversity in North American pines. *Am. J. Bot.* **1988**, *75*, 353–368. [CrossRef]

26. Arno, S.F. Community types and natural disturbance processes. In *Whitebark Pine Communities: Ecology and Restoration*; Tomback, D.F., Arno, S.F., Keane, R.E., Eds.; Island Press: Washington, WA, USA, 2001; pp. 74–88.

27. Whitebark Pine Ecosystem Foundation. Rangewide Map for Whitebark Pine. Whitebark Pine Ecosystem Foundation: 2015. Available online: http://Whitebarkfound.org/wp-content/uploads/2014/05/whitebark-pine-range-colour-2014.jpg (accessed on 10 August 2015).

28. Tomback, D.F.; Arno, S.F.; Keane, R.E. The compelling case for management intervention. In *Whitebark Pine Communities: Ecology and Restoration*; Tomback, D.F., Arno, S.F., Keane, R.E., Eds.; Island Press: Washington, DC, USA, 2001; pp. 3–25.

29. Dayton, P.K. Towards an Understanding of Community Resilience and the Potential Effects of Enrichments to the Benthos at McMurdo Sound, Antarctica. In *Proceedings of the Colloquium on Conservation Problems in Antarctica*; Parker, B.C., Ed.; Allen Press: Lawrence, KS, USA, 1972; pp. 81–96.

30. Ellison, A.M.; Bank, M.S.; Clinton, B.D.; Colburn, E.A.; Elliott, K.; Ford, C.R.; Foster, D.R.; Kloeppel, B.D.; Knoepp, J.D.; Lovett, G.M.; *et al.* Loss of foundation species: Consequences for the structure and dynamics of forested ecosystems. *Front. Ecol. Environ.* **2005**, *3*, 479–486. [CrossRef]

31. Mills, L.S.; Soulé, M.E.; Doak, D.F. The keystone-species concept in ecology and conservation. *BioScience* **1993**, *43*, 219–224. [CrossRef]

32. Soulé, M.E.; Estes, J.A.; Berger, J.; Martinez del Rio, C. Ecological effectiveness: Conservation goals for interactive species. *Conserv. Biol.* **2003**, *17*, 1238–1250. [CrossRef]

33. Costanza, R.; d'Arge, R.; de Groot, R.; Farber, S.; Grasso, M.; Hannon, B.; Limburg, K.; Naeem, S.; O'Neill, R.V.; Paruelo, J.; *et al.* The value of the world's ecosystem services and natural capital. *Nature* **1997**, *387*, 253–260. [CrossRef]

34. Millenium Ecosystem Assessment. *Ecosystem and Human Well-being: Synthesis*; Island Press: Washington, DC, USA, 2005.

35. Farnes, P.E. SNOTEL and Snow Course Data: Describing the Hydrology of Whitebark Pine Ecosystems. In *Proceedings—Symposium on Whitebark Pine Ecosystems: Ecology and Management of a High-Mountain Resource*; Schmidt, W.C., McDonald, K.J., Eds.; General Technical Report INT-270; USDA Forest Service, Intermountain Research Station: Ogden, UT, USA, 1990; pp. 302–304.

36. Callaway, R.M. Competition and facilitation on elevation gradients in subalpine forests of the northern Rocky Mountains, USA. *Oikos* **1998**, *2*, 561–573. [CrossRef]

37. Tomback, D.F.; Kendall, K.C. Biodiversity losses: The downward spiral. In *Whitebark Pine Communities: Ecology and Restoration*; Tomback, D.F., Arno, S.F., Keane, R.E., Eds.; Island Press: Washington, DC, USA, 2001; pp. 243–262.

38. Resler, L.M.; Tomback, D.F. Blister rust prevalence in krummholz whitebark pine: Implications for treeline dynamics. *Arct. Antarct. Alp. Res.* **2008**, *40*, 161–170. [CrossRef]

39. Moerman, D.E. *Native American Ethnobotany*; Timber Press: Portland, OR, USA, 1998.

40. Moerman, D.E. *Native American Medicinal Plants: An Ethnobotany dictionary*; Timber Press: Portland, OR, USA, 2009.

41. Marsh, P.; Quinton, B.; Pomeroy, J. Hydrological Processes and Runoff at the Arctic Treeline in Northwestern Canada. In Proceedings of the Tenth International Northern Research Basins Symposium and Workshop, Norway, 1994; Sand, K., Killingtveit, A., Eds.; SINTF: Trondheim, Norway, 1995; pp. 368–397.

42. Hiemstra, C.A.; Liston, G.E.; Reiners, W.A. Snow redistribution by wind and interactions with vegetation at upper treeline in the Medicine Bow Mountains, Wyoming, USA. *Arct. Antarct. Alp. Res.* **2002**, *34*, 262–273. [CrossRef]

43. Barnett, T.P.; Adam, J.C.; Lettenmaier, D.P. Potential impacts of a warming climate on water availability in snow-dominated regions. *Nature* **2005**, *438*, 303–309. [CrossRef] [PubMed]

44. Viviroli, D.; Dürr, H.H.; Messerli, B.; Meybeck, M. Mountains of the world, water towers for humanity, typology, mapping, and global significance. *Water Resour. Res.* **2007**, *43*, W07447. [CrossRef]

45. Tomback, D.F.; Chipman, K.G.; Resler, L.M.; Smith-McKenna, E.K.; Smith, C.M. Relative abundance and functional role of whitebark pine at treeline in the northern Rocky Mountains. *Arct. Antarct. Alp. Res.* **2014**, *46*, 407–418. [CrossRef]

46. Blakeslee, S.C. *Assessing Whitebark Pine Vigor and Facilitation Roles in the Alpine Treeline Ecotone*; Masters of Science, University of Colorado Denver: Denver, CO, USA, 2012.

47. Pyatt, J.C.; Tomback, D.F.; Blakeslee, S.C.; Wunder, M.B.; Resler, L.M.; Boggs, L.A.; Bevency, H.D. The importance of conifers for facilitation at treeline: Comparing biophysical characteristics of leeward microsites in whitebark pine communities. *Arct. Antarct. Alp. Res.* **2016**. in review.

48. Smith, E.K.; Resler, L.M.; Vance, E.A.; Carstensen, L.W., Jr.; Kolivras, K.N. Blister rust incidence in treeline whitebark pine, Glacier National Park, USA: Environmental and topographic influences. *Arct. Antarct. Alp. Res.* **2011**, *43*, 107–117. [CrossRef]

49. Smith-McKenna, E.K.; Resler, L.M.; Tomback, D.F.; Zhang, H.; Malanson, G.P. Topographic influences on the distribution of white pine blister rust in *Pinus albicaulis* treeline communities. *Écoscience* **2013**, *20*, 215–229. [CrossRef]

50. R Core Team. R: A language and environment for statistical computing. R Foundation for Statistical Computing: Vienna, Austria, 2015; Version 3.22. Available online: https://www.R-project.org/ (accessed on 2 October 2015).

51. Tomback, D.F.; Resler, L.M. Structure and composition of Rocky Mountain treeline whitebark pine communities at their northern boundary. University of Colorado Denver: Denver, CO, USA, 2015; in preparation.

52. Resler, L.M.; Shao, Y.; Tomback, D.F.; Malanson, G.P. Predicting the functional role and occurrence of whitebark pine (*Pinus albicaulis*) at alpine treeline: Model accuracy and variable importance. *Ann. Assoc. Am. Geogr.* **2014**. [CrossRef]

53. Resler, L.M; Tomback, D.F.; Malanson, G.P. Treeline community structure and composition on Parker Ridge, Banff National Park. Virginia Tech: Blacksburg, VA, USA, unpublished data; 2015.

54. Wagner, A.C.; Tomback, D.F.; Pansing, E.R. Structure and composition of treeline communities at whitebark pine's (*Pinus albicaulis*) southern distributional limit in the Rocky Mountains. University of Colorado: Denver, CO, USA, 2015; in preparation.

55. Resler, L.M.; Shao, Y. Assessing whitebark pine treeline communities and blister rust infection in Grand Teton National Park. Virginia Tech: Blacksburg, VA, USA, unpublished data; 2015.

56. Arno, S.F.; Hammerly, R.P. *Timberline: Mountain and Arctic Forest Frontiers*; Timber Press: Portland, OR, USA, 1984.

57. Finklin, A. *A Climate handbook for Glacier National Park—With data for Waterton Lakes National Park*; General Technical Report GTR INT-204; USDA Forest Service, Intermountain Research Station: Ogden, UT, USA, 1986.

58. Mellor, M. Blowing snow. *USA Cold Reg. Res. Eng. Lab. Monogr.* **1965**, *III-A3c*, 1–52.

59. Finney, E.A. Snow control on the highways. In *Michigan Engineering Experimental Station, Bulletin*; Michigan State College: Lansing, MI, USA, 1934; Volume 75, pp. 1–81.

60. Calkins, D.J. Simulated snowdrift patterns. *USA Cold Reg. Res. Eng. Lab. Spec. Rep.* **1975**, *219*, 1–16.

61. Kind, R.J. Snow Drifting. In *Handbook of Snow*; Gray, D.M., Male, D.H., Eds.; Pergamon Press: New York, NY, USA, 1981; pp. 338–358.

62. Daly, C. Snow distribution patterns in the alpine krummholz zone. *Prog. Phys. Geogr.* **1984**, *8*, 157–175. [CrossRef]

63. Allen, J.R.L. *Current Ripples*; North Holland Publishing Company: Amsterdam, Netherlands, 1968.

64. Kobayashi, D. Studies of snow transport in low-level drifting snow. *Inst. Low Temp. Sci. Rep.* **1972**, *231*, 1–58.

65. Sturges, D.L. Snow fencing to increase streamflow: Preliminary results. *West. Snow Confer. Proc.* **1986**, *54*, 18–28. Available online: http://www.westernsnowconference.org/sites/westernsnowconference.org/PDFs/1986Sturges.pdf (accessed on 21 December 2015).

66. Geddes, C.A.; Brown, D.G.; Fagre, D.B. Topography and vegetation as predictors of snow water equivalent across the Alpine Treeline Ecotone at Lee Ridge, Glacier National Park, Montana, USA. *Arct. Alp. Res.* **2005**, *37*, 19–205.

67. Vajda, A.; Venäläinen, A.; Hänninen, P.; Sutinen, R. Effect of vegetation on snow cover at the northern timberline: A case study in Finnish Lapland. *Silva Fenn.* **2006**, *40*, 195–207. [CrossRef]

68. Liptzin, D.; Seastedt, T. Patterns of snow, deposition, and soil nutrients at multiple spatial scales at a Rocky Mountain tree line ecotone. *J. Geophys. Res.* **2009**, *114*, G04002. [CrossRef]

69. Jonas, T.; Marty, C.; Magnusson, J. Estimating the snow water equivalent from snow depth measurements in the Swiss Alps. *J. Hydrol.* **2009**, *378*, 161–167. [CrossRef]

70. Anderton, S.P.; White, S.M.; Alvera, B. Evaluation of spatial variability in snow water equivalent for a high mountain catchment. *Hydrol. Process.* **2004**, *19*, 435–453. [CrossRef]

71. Pomeroy, J.W.; Brun, E. Physical Properties of Snow. In *Snow Ecology: An Interdisciplinary Examination of Snow-Covered Ecosystems*; Jones, H.G., Pomeroy, J.W., Walker, D.A., Hoham, R.W., Eds.; Cambria University Press: Cambridge, UK, 2001; pp. 45–126.

72. Tyler, S.; Burak, S.A.; McNamara, J.P.; Lamontagne, A.; Selker, J.S.; Dozier, J. Spatially distributed temperatures at the base of two mountain snowpacks measured with fiber-optic sensors. *J. Glaciol.* **2008**, *54*, 673–679. [CrossRef]

73. Billings, W.D.; Bliss, L.C. An alpine snowbank environment and its effects on vegetation, plant development, and productivity. *Ecology* **1959**, *40*, 388–397. [CrossRef]

74. Billings, W.D. Vegetational pattern near alpine timberline as affected by fire-snowdrift interactions. *Vegetatio* **1969**, *19*, 192–207. [CrossRef]

75. Walker, D.A.; Halfpenny, J.C.; Walker, M.D.; Wessman, C.A. Long-term studies of snow-vegetation interactions. *BioScience* **1993**, *43*, 287–301. [CrossRef]

76. Odland, A.; Munkejord, H.K. Plants as indicators of snow layer duration in southern Norwegian mountains. *Ecol. Indic.* **2008**, *8*, 57–68. [CrossRef]

77. Fahnestock, J.; Povrik, K.; Welker, J. Ecological significance of litter redistribution by wind and snow in arctic landscapes. *Ecography* **2008**, *23*, 623–631. [CrossRef]

78. Nardi, A.; Bowman, W. Hot spots of inorganic nitrogen availability in an alpine-subalpine ecosystem, Colorado Front Range. *Ecosystems* **2011**, *14*, 848–863. [CrossRef]

79. Tomback, D.F.; Resler, L.M. Invasive pathogens at alpine treeline: Consequences for treeline dynamics. *Phys. Geogr.* **2007**, *28*, 397–418. [CrossRef]

80. Hiemstra, C.A.; Liston, G.E.; Reiners, W.A. Observing, modelling, and validating snow redistribution by wind in a Wyoming upper treeline landscape. *Ecol. Model.* **2006**, *197*, 35–51. [CrossRef]

81. Smith, C.M.; Wilson, B.; Rasheed, S.; Walker, R.C.; Carolin, T.; Shepherd, R. Whitebark pine and white pine blister rust in the Rocky Mountains of Canada and northern Montana. *Can. J. For. Res.* **2008**, *38*, 982–985. [CrossRef]

82. Keane, R.E.; Tomback, D.F.; Aubry, C.A.; Bower, A.D.; Campbell, E.M.; Cripps, C.L.; Jenkins, M.B.; Mahalovich, M.F.; Manning, M.; McKinney, S.T.; *et al. A Range-Wide Restoration Strategy for Whitebark Pine (Pinus albicaulis)*; General Technical Report RMRS-GTR-279; USDA Forest Service: Fort Collins, CO, USA, 2012.

83. GYWPMWG (Greater Yellowstone Whitebark Pine Monitoring Working Group). *Monitoring Whitebark Pine in the Greater Yellowstone Ecosystem*; 2012 Annual Report, Natural Resource Data Series NPS/GRYN/NRDS—2013/498; Chambers, N., Ed.; U.S. Department of the Interior, National Park Service, Natural Resource Stewardship and Science: Fort Collins, CO, USA, 2013.

84. Smith, C.M.; Shepherd, B.; Gillies, C.; Stuart-Smith, J. Changes in blister rust infection and mortality in whitebark pine over time. *Can. J. For. Res.* **2013**, *43*, 90–96. [CrossRef]

85. McDonald, G.I.; Hoff, R.J. Blister rust: An introduced plague. In *Whitebark Pine Communities: Ecology and Restoration*; Tomback, D.F., Arno, S.F., Keane, R.E., Eds.; Island Press: Washington, DC, USA, 2001; pp. 193–220.

86. Geils, B.W.; Hummer, K.E.; Hunt, R.S. White pines, *Ribes*, and blister rust: A review and synthesis. *For. Pathol.* **2010**, *40*, 147–185. [CrossRef]

87. Tomback, D.F. Post-fire regeneration of krummholz whitebark pine: A consequence of nutcracker seed caching. *Madroño* **1986**, *33*, 100–110.

88. McKinney, S.T.; Fiedler, C.E.; Tomback, D.F. Invasive pathogen threatens bird-pine mutualism: Implications for sustaining a high-elevation ecosystem. *Ecol. Appl.* **2009**, *19*, 597–607. [CrossRef] [PubMed]

89. Barringer, L.; Tomback, D.F.; Wunder, M.B.; McKinney, S.T. Whitebark pine stand condition, tree abundance, and cone production as predictors of visitation by Clark's Nutcracker. *PLoS ONE* **2012**, *7*, e37663. [CrossRef] [PubMed]

90. Malanson, G.P.; Butler, D.R.; Fagre, D.B.; Walsh, S.J.; Tomback, D.F.; Daniels, L.D.; Resler, L.M.; Smith, W.K.; Weiss, D.J.; Peterson, D.L.; *et al.* Alpine treeline of western North America: Linking organism-to-landscape dynamics. *Phys. Geogr.* **2007**, *28*, 378–396. [CrossRef]

91. Smith-McKenna, E.K.; Malanson, G.P.; Resler, L.M.; Carstensen, L.W.; Prisley, S.P.; Tomback, D.F. Cascading effects of feedbacks, disease, and climate change on alpine treeline dynamics. *Environ. Model. Softw.* **2014**, *62*, 85–96. [CrossRef]

92. Primack, R.B. *Essentials of Conservation Biology*, 6th ed.; Sinauer: Sunderland, MA, USA, 2014.

93. Environment Canada. Recovery Strategy for Whitebark Pine (*Pinus albicaulis*) in Canada [Draft]. In *Species at Risk Act Recovery Strategy Series*; Environment Canada: Ottawa, ON, Canada, 2015.

94. Keane, R.E.; Holsinger, L.; Mahalovich, M.F.; Tomback, D.F. *Restoring Whitebark Pine in the Face of Climate Change*; General Technical Report RMRS-GTR-XXX; USDA Forest Service, Rocky Mountain Research Station: Fort Collins, CO, USA, 2016; in press.

95. Keane, R.E.; Parsons, R. Restoring whitebark pine forests of the Northern Rocky Mountains, USA. *Ecol. Restor.* **2010**, *28*, 56–70. [CrossRef]

96. Pansing, E.R.; Tomback, D.F.; Wunder, M.B.; French, J.P.; Wagner, A.C. Simulated seed dispersal reveals geographic and community-based patterns of seed pilferage, germination, and seedling survival in whitebark pine (*Pinus albicaulis*). *Oecologia* **2016**, submitted.

forests

MDPI

Review

Building on Two Decades of Ecosystem Management and Biodiversity Conservation under the Northwest Forest Plan, USA

Dominick A. DellaSala [1,*], **Rowan Baker** [2], **Doug Heiken** [3], **Chris A. Frissell** [4], **James R. Karr** [5], **S. Kim Nelson** [6], **Barry R. Noon** [7], **David Olson** [8] and **James Strittholt** [9]

[1]   Geos Institute, 84-4th Street, Ashland, OR 97520, USA
[2]   Independent Consultant, 2879 Southeast Kelly Street, Portland, OR 97202, USA;
      watershedfishbio@yahoo.com
[3]   Oregon Wild, P.O. Box 11648, Eugene, OR 97440, USA; dh@oregonwild.org
[4]   Flathead Lake Biological Station, 32125 Bio Station Lane, University of Montana, Polson,
      MT 59860-6815 USA; leakinmywaders@yahoo.com
[5]   102 Galaxy View Court, Sequim, WA 98382, USA; jrkarr@u.washington.edu
[6]   Department of Fisheries and Wildlife, Oregon State University, 104 Nash Hall,
      Corvallis, OR 97331-3803, USA; kim.nelson@oregonstate.edu
[7]   Department of Fish, Wildlife and Conservation Biology, Colorado State University,
      Fort Collins, CO 80523, USA; Barry.Noon@colostate.edu
[8]   Conservation Earth Consulting; 4234 McFarlane Avenue, Burbank, CA 91505, USA;
      conservationearth@live.com
[9]   Conservation Biology Institute, 136 SW Washington Ave #202, Corvallis, OR 97333, USA; stritt@conbio.org
*   Author to whom correspondence should be addressed; dominick@geosinstitute.org;
      Tel.: +541-482-4459 (ext. 302); Fax: +1-541-482-4878.

Academic Editor: Diana F. Tomback
Received: 20 June 2015; Accepted: 14 September 2015; Published: 22 September 2015

Abstract: The 1994 Northwest Forest Plan (NWFP) shifted federal lands management from a focus on timber production to ecosystem management and biodiversity conservation. The plan established a network of conservation reserves and an ecosystem management strategy on ~10 million hectares from northern California to Washington State, USA, within the range of the federally threatened northern spotted owl (*Strix occidentalis caurina*). Several subsequent assessments—and 20 years of data from monitoring programs established under the plan—have demonstrated the effectiveness of this reserve network and ecosystem management approach in making progress toward attaining many of the plan's conservation and ecosystem management goals. This paper (1) showcases the fundamental conservation biology and ecosystem management principles underpinning the NWFP as a case study for managers interested in large-landscape conservation; and (2) recommends improvements to the plan's strategy in response to unprecedented climate change and land-use threats. Twenty years into plan implementation, however, the U.S. Forest Service and Bureau of Land Management, under pressure for increased timber harvest, are retreating from conservation measures. We believe that federal agencies should instead build on the NWFP to ensure continuing success in the Pacific Northwest. We urge federal land managers to (1) protect all remaining late-successional/old-growth forests; (2) identify climate refugia for at-risk species; (3) maintain or increase stream buffers and landscape connectivity; (4) decommission and repair failing roads to improve water quality; (5) reduce fire risk in fire-prone tree plantations; and (6) prevent logging after fires in areas of high conservation value. In many respects, the NWFP is instructive for managers considering similar large-scale conservation efforts.

Forests **2015**, *6*, 3326–3352

Keywords: biodiversity; climate change; ecological integrity; ecosystem management; global forest model; Northwest Forest Plan; northern spotted owl

---

1. Introduction

The 1994 Northwest Forest Plan (NWFP) ushered in ecosystem management and biodiversity conservation on nearly 10 million ha of federal lands within the range of the federally threatened northern spotted owl (*Strix occidentalis caurina*) from northern California to Washington State, mostly along the western slopes of the Cascade Mountains, USA (Figure 1). The plan was prepared in response to a region wide legal injunction on logging of spotted owl habitat (older forests) issued in 1991 by U.S. District Court Judge William Dwyer. After reviewing the NWFP, Judge Dwyer ruled that the plan was the *"bare minimum"* (emphasis added) necessary for the Bureau of Land Management (BLM) and the U.S. Forest Service to comply with relevant statutes (see http://www.justice.gov/enrd/3258.htm; accessed on 29 July 2015). The plan's conservation framework and unprecedented monitoring of forest and aquatic conditions along with at-risk species (those with declining populations) offer important lessons for managers interested in large-scale conservation and ecosystem management [1]. Thus, our objectives are to: (1) showcase the plan's fundamental conservation biology and ecosystem management principles as a regional case study for large-scale forest planning; and (2) build on the plan's conservation approach to provide a robust strategy for forest biodiversity in the context of unprecedented climate change, increasing land-use stressors, and new forest and climate science and policies.

At the time of the NWFP development, President Bill Clinton sought to end decades of conflict over old-growth logging by directing 10 federal agencies responsible for forest management, fisheries, wildlife, tribal relations, and national parks to work together and with scientists on a region wide forest plan that would be "scientifically sound, ecologically credible, and legally responsible." The plan was crafted to ensure the long-term viability of "our forests, our wildlife, and our waterways," and to "produce a predictable and sustainable level of timber sales and non-timber resources that will not degrade or destroy the environment." A multi-disciplinary team of scientists known as the Forest Ecosystem Management Assessment Team [2] was tasked with identifying management alternatives that would meet the requirements of applicable laws and regulations, including the Endangered Species Act, the National Forest Management Act, the Federal Land Policy Management Act, the Clean Water Act, and the National Environmental Policy Act.

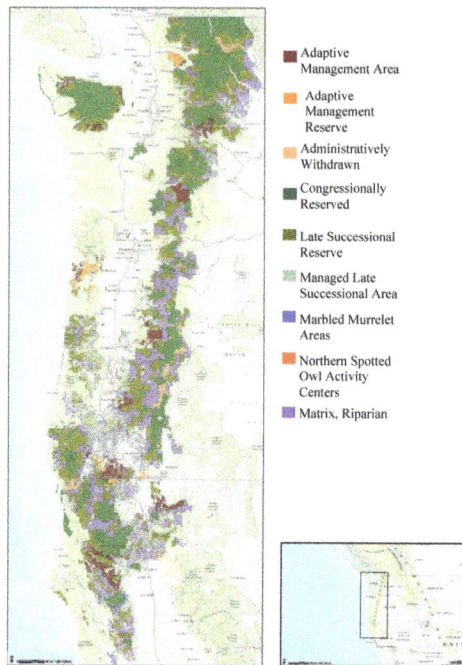

Figure 1. Land-use allocations within the Northwest Forest Plan (NWFP) area: Congressionally reserved—2.93 million ha (30%); Late Successional Reserves (LSRs)—2.96 million ha (30%); Managed Late Successional Reserves—40,880 ha (1%); Adaptive Management Areas—608,720 ha (6%); Administratively Withdrawn 590,840 ha (6%); Riparian Reserves—1.1 million ha (11%); and Matrix—1.6 million ha (16%). Figure created using Data Basin (www.databasin.org; accessed on 29 July 2015) and NWFP data layers [3].

The NWFP amended resource management plans for 19 national forests and seven BLM planning districts with 80% of those lands dedicated to some form of conservation (Figure 1). This increased level of protection and improved management standards were necessary because for many decades federal lands were managed without proper regard for water quality, fish and wildlife viability, and ecosystem integrity. Overcutting of older forests and rapid road expansion were the main factors responsible for the 1990 threatened species listing of the northern spotted owl, 1992 threatened listing of the marbled murrelet (*Brachyramphus marmoratus*), multiple listings of Evolutionary Significant Units (ESUs) of salmonids (*Oncorhynchus* spp.), and pervasive and mounting water quality problems. Prior to the NWFP, ~9.6 million cubic meters of timber was being logged from old-growth forests (>150 years old) annually on federal lands alone—roughly 5 square kilometers per week (assuming stands averaged 300 cubic meters per hectare). USFWS [4] estimated that this rate of logging would have eliminated spotted owl habitat outside remote and protected areas within a few decades. Simultaneously, logging was on the brink of eliminating old-growth forests from surrounding nonfederal lands.

Older forests in the Pacific Northwest are a conservation priority because they harbor exceptional levels of forest biodiversity (e.g., >1000 species have been recognized) and numerous at-risk species [2]. Historically, such forests widely dominated much of the Pacific Northwest landscape, especially in wet areas (coastal) where the intervals between successive fires were centuries long [5].

Older forest communities vary considerably in dominant tree species composition among the southern Cascade Range (Oregon/California), central and northern Cascades (Oregon/Washington), Coast Range (California/Oregon/Washington) and Klamath Mountains (Oregon/California [6]).

Forests **2015**, *6*, 3326–3352

Forests are generally dominated by Douglas-fir (*Pseudotsuga menziesii*) on sites associated with western hemlock (*Tsuga heterophylla*, sometimes including Pacific and grand fir, *Abies amabalis*, *A. grandis*; western red cedar *Thuja plicata*, bigleaf maple *Acer macrophyllum*); mixed conifers (white fir *A. concolor* and sometimes incense cedar *Calocedrus decurrens*, ponderosa and sugar pine *Pinus ponderosa*, *P. lambertina*); and mixed-evergreens (Pacific madrone *Arbutus menziesii*, tan oak *Lithocarpus densiflorus*, and canyon live oak *Quercus chrysolepis*). Structurally, these forests are characterized by the presence of high densities of large (>100 cm in diameter) conifers (typically 16–23 trees/ha), varied tree sizes and multi-layered canopies, trees with broken and dead tops, high levels of snags and downed wood, and diverse understories [6].

Most forest types in this region generally begin acquiring older forest characteristics at 80 years, depending on site productivity and disturbance history, with full expression of structural diversity at 400+ years [7]. Upper elevation subalpine fir (*Abies lasiocarpa*) and Pacific silver fir are not considered old growth until they are 260–360 years old [8]. Notably, researchers have recently developed an old-growth structure index (OGSI) to represent a successional continuum from young to older forests. The OGSI is a continuous value of 0–100 used to delineate older forests based on four features: (1) large live tree density; (2) large snag density; (3) down wood cover; and (4) tree size diversity at the stand level [9]. Young forests <80 years old that originate from natural disturbance in older forests, known as complex early seral forest, also have high levels of structural complexity (e.g., snags and downed logs) and species richness (especially forbs, shrubs; [10,11]). These younger forests have only recently been recognized as a conservation priority and like old growth have been replaced by structurally simplistic tree plantations [10].

2. NWFP's Long-Term Objectives

FEMAT [2] aptly recognized that even with the plan's protective elements in place, it would take at least a century and possibly two to restore a functional, interconnected late-successional/old growth (LSOG) ecosystem because older forests were reduced to a fraction (<20%) of their historical extent, and 40% of the LSRs were regenerating from prior clearcut harvest that would require decades of restoration to eventually acquire older characteristics [12]. The NWFP also represented a tradeoff between conservation and timber interests with about 1.6 million ha (16%) of older forests placed into the "Matrix" (Figure 1) where the majority of logging would take place pursuant to the plan's management standards and guidelines. As the NWFP was implemented, the volume of timber anticipated for sale (known as the probable sale quantity) was projected at ~234 million cubic meters annually. Since then, the plan has achieved about 80% of the probable sale quantity (on average ~178 million m^3 annually [13]). The apparent shortfall has been variously attributed to protective measures implemented before timber volume can be offered for sale, ongoing public controversy (appeals and lawsuits) around logging of older forests in the Matrix, fluctuations in domestic housing starts and global timber markets. Congressional appropriations to federal agencies for administering timber sales also have contributed to a *de facto* limit on timber offered for sale. Consequently, the plan's timber goals remain controversial. Some contend that socioeconomic considerations tied to timber extraction have not been met [14]. Others contend that rural communities no longer depend on timber in a region where economic sectors are influenced mainly by external factors and local economies have largely diversified [15]. Nonetheless, while it is premature to judge the efficacy of a 100-year plan in just two decades, periodic monitoring has shown that it has put federal forestlands on a trajectory to meet many of its ecosystem management targets [1,9,16,17].

Restoring a functional, interconnected LSOG ecosystem requires protecting existing older forests and growing more of it over time from young-growth tree plantations within the reserves. Restoring LSOG from former tree plantations is an uncertain endeavor that will require many decades to centuries and has never been envisioned before on such a large scale, especially in the face of rapidly changing climate. Thus, periodic monitoring of several of the ecosystem-based components of the NWFP by federal agencies is being used to gauge restoration targets, assess implementation efficacy of the plan, and proactively respond to new stressors. For instance, an unprecedented level of old forest, aquatics, and at-risk species monitoring occurs at regular intervals, depending on factors assessed, in order to achieve compliance with the 1991 Dwyer court ruling and biodiversity requirements of the National Forest Management Act of 1976. Maintaining biodiversity is a fundamental goal of any large conservation effort and the NWFP is instructive for managers considering similar large-scale ecosystem management and conservation efforts.

2.1. Reserves as a Coarse Filter

Conservation scientists have long-recognized that effective conservation planning involves two complementary approaches: a coarse filter consisting of representative reserve networks, and fine filter that includes local protections for species outside reserves [18,19]. FEMAT [2] emphasized the need for a large, interconnected reserve network as fundamental to biodiversity conservation [1,20,21] (IUCN protected areas categories: http://www.iucn.org/about/work/programmes/gpap_home/gpap_ quality/gpap_pacategories/; accessed on 17 September 2015). Thus, the conservation foundation of the NWFP is rooted in a network of reserves (e.g., LSRs and Riparian Reserves) that are widely distributed (Figure 1) throughout the planning area. The reserve network was principally designed to support viability and dispersal of the northern spotted owl in what is otherwise a highly fragmented system (Figure 2).

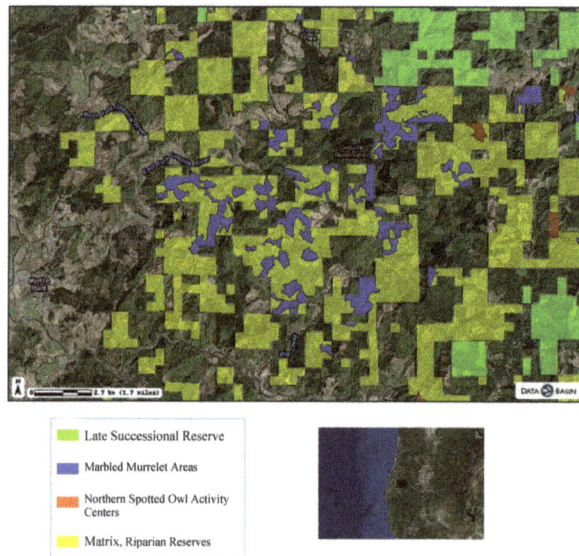

Figure 2. Satellite image of Southwest Oregon showing extensive fragmentation from a "checkerboard" pattern of clearcuts on private and public lands with NWFP land management allocations. Map created using Data Basin (www.databasin.org; accessed on 15 September 2015).

With reserves acting as a coarse filter, ecosystem-based approaches can be implemented to target geographic concentrations—or hotspots—of listed or rare species, thereby increasing conservation efficacy via multiple species benefits. Coarse filters are landscape characteristics of a natural environment that are easily measured, for instance, using satellite images, digital elevation models, and weather station data. Importantly, coarse filters are meant to capture the habitat needs of an entire species assemblage rather than habitat requirements for a particular focal species. For example, a land manager might use dominant vegetation identified through remotely sensed imagery to infer which species potentially occur across the landscape. Thus, the fundamental premise of coarse filters is that measuring the amounts and spatial distribution of biophysical features allows managers to assess the suitability of the landscape for multiple species and to represent key aggregate ecological attributes within a system of designated reserves. Effective coarse-filter reserves need to be defined at appropriate scales so that habitats and populations are sufficiently represented and reserves are distributed in redundant sequences to be robust to prevailing dynamics of natural biophysical disturbance (e.g., forest fires) and external land-management stressors in the surrounding landscape. These considerations were explicitly implemented by FEMAT when scientists designed alternatives that established the conservation architecture of the NWFP.

Three scales are important for estimating the amount and spatial arrangement of habitat needed to recover or conserve at-risk species, particularly those that are indicators of a broader community:

(1) **Species**: habitat needed to provide the resources and physical conditions required for a particular species to survive and reproduce.
(2) **Population**: habitat needed to support a local population of sufficient size to be resilient to background stochastic demographic and environmental events and short-term inbreeding depression.
(3) **Geographic range**: collective habitat required by multiple local populations of a species that are well distributed so that all populations do not respond synchronously to stochastic environmental events.

Central to its biodiversity focus, the NWFP was designed with explicit consideration of resilience, redundancy, and representation across multiple groups of taxa and communities. Resilient populations are those that are large enough, have sufficient genetic variation, and are sufficiently diverse with respect to the age and sex of individuals to persist in the face of periodic threats such as drought, wildfire, disease, and climate change. With respect to redundancy in populations or habitat areas, sufficient numbers of separate populations of a species and areas to support them are needed to provide a margin of safety in case disturbance eliminates some populations or important habitat types. In addition, sufficient genetic variation among populations of a species is necessary to conserve the breadth of the species' genetic makeup and its capacity to evolve and adapt to new environmental conditions. Representation refers to the plan's ability to capture a range of old growth conditions regionally within a reserve network.

2.2. Survey and Manage Program as Fine Filter

As a supplement to the Endangered Species Act, one of the fine-filters of the NWFP is the "survey and manage" program, an unprecedented precautionary approach designed to protect known locations and collect new information to address persistence probabilities and management uncertainties for rare and poorly surveyed species outside the reserve network [22]. Some 400 late-successional species of amphibians, bryophytes, fungi, lichens, mollusks, vascular plants, arthropod functional groups, and one mammal, including many endemics that otherwise may not persist outside the reserve network, were included in the program and given limited protections from logging if found (usually small site-specific buffers).

The survey and manage standards and guidelines for management might not be needed if the coarse filter reserves and older forests were fully functional and, therefore, resilient to short-term disturbance like fires and longer-term climate and land-use changes. However, that is not currently the case. In sum, the survey and manage program resulted in significant gains in knowledge, reduced uncertainty about conservation, and developed useful new inventory methods for rare species [22]. The program, however, remains one of the more controversial aspects of the NWFP, and federal agencies have repeatedly proposed its elimination given the restrictions it can place on the pace and cost of logging.

Thorough documentation of old forest species' distributions and diversity is still needed. In particular, some regions with diverse vegetation types (e.g., Klamath-Siskiyou of southwest

Oregon/northern California [23]) have exceptional concentrations of endemic species that remain poorly studied and vulnerable to climate change [24]. Many rare species are inadequately known for development of effective management policies and practices, especially under a rapidly changing climate. The survey and manage program is also needed to ensure that rare species do not become at-risk species due to unforeseen population declines and conservation neglect.

2.3. Northern Spotted Owl Decline Slowed but Not Reversed

Spotted Owl Conservation Strategy—The northern spotted owl is the umbrella species for hundreds of late-successional species in the NWFP area [2]. When developing the conservation strategy for the owl, Thomas *et al.* [25] drew on fundamental principles from population viability analysis [26], island biogeography [27], and conservation biology [28–30] that applied both specifically to the owl and more generally to the community of late-successional associates. Thus, the NWFP is considered a model for conserving at-risk species [1]. Additional conservation biology principles guided the design of the NWFP [2]:

- Species that are widely distributed are less prone to extinction than those with more restricted ranges because local population dynamics are more independent [31].
- Large patches of habitat supporting many individuals are more likely to sustain those populations than small patches because larger populations are less subject to demographic and environmental stochasticity [32,33].
- Populations residing in habitat patches in close proximity are less extinction prone than those in widely separated patches because the processes of dispersal and recolonization are facilitated [34].
- The extent to which the landscape matrix among habitat patches (supporting local populations of the focal species) resembles suitable habitat, the greater the connectivity among local populations leading to lower extinction risks [35].
- Sustaining a species over the long-term requires that demographic processes be evaluated at three key spatial scales: territory, local population, and metapopulation [36].

Spotted Owl Population Trends and the NWFP—Even with the reserve network in place, spotted owl populations on federal lands have continued to show an alarming (3.8%) annual rate of decline [9] that has increased from the 2.8% annual decline reported previously [37]. Spotted owl populations are monitored across 11 large demographic study areas on federal (n = 8) and nonfederal (n = 3) lands where data on owl population dynamics are collected. Based on 2011 monitoring results for demography study areas, four study areas showed marked declines (both the point estimator and 95% confidence intervals) in mean annual rate of owl population change [38]. In 2015, the number of study areas with marked declines in owl populations increased to six (K.M. Dugger, pers. communication). Spotted owl declines were attributed to interference competition with barred owls (*Strix varia*; [39]), logging-related habitat losses (mostly nonfederal lands), and the lack of a fully functional reserve system [12,40].

Notably, total spotted owl detections and the number of previously banded owls was the lowest ever recorded for the demography study areas [41]. Spotted owl detections at historic territories remained unchanged from 2013–2014 at LSRs, whereas, a double-digit decrease in owl detections was noted in the Matrix that well exceeded the slight decrease in detections recorded for Wilderness areas. Anthony *et al.* [42] also reported that the decline in spotted owls was steepest on study areas not managed under the NWFP and therefore the downward trajectory of owl populations might have been much worse without the NWFP.

Spotted Owl Habitat Trends—Before the NWFP, the annual rate of LSOG losses on national forests was ~1% in California and 1.5% in Oregon and Washington [9,40]. Recent monitoring of older forests by federal agencies using multiple inventory methods shows, at the forest plan-scale, a slight reduction in the area of federal older forests (2.8%–2.9% in 2012 compared to 1993 levels Table 1).

Table 1. Total old forest area (hectares x million) for federal (USFS, BLM combined) *vs.* nonfederal lands using three old-forest estimates: an old-growth structure index at 80-years (OGSI-80); old-growth structure index at 200 years (OGSI-200); and Late-Successional/Old Growth (LSOG) [9]. Percent differences between time periods (parentheses) were repeated from Davis *et al.* [9] who used more significant figures in calculations not shown here and rounded to the nearest hundred thousand.

Time Period	Federal OGSI-80	Federal OGSI-200	Federal LSOG	NonFederal OGSI-80	NonFederal OGSI-200	NonFederal LSOG
1993	5.1	2.6	3.0	2.6	0.7	1.6
2012	4.9 (−2.9)	2.5 (−2.8)	2.6 (−2.0)	2.3 (−11.6)	0.6 (−18.1)	1.3 (−14.2)

Based on federal lands monitoring reports, wildfire accounted for 4.2%–5.4% of the gross older forest losses compared to logging, which accounted for 1.2%–1.3% old-forest reductions [9]. Such losses were within the 5% anticipated disturbance level for the NWFP area over this time frame; however, fire-related losses were >5% in some dry forest ecoprovinces (5.5%–7.1% Washington Eastern Cascades; 12.2%–15.3% Klamath Oregon; and 7.0%–13.1% California) [9]. Thus, one primary accomplishment of the plan was to drastically slow old forest losses from logging over the NWFP time period. Exceptions include BLM lands in western Oregon, where the rate of old forest loss was >2 times that of U.S. Forest Service lands over a 10-year period (Table 2).

Table 2. Estimated spotted owl habitat losses due to logging on U.S. Forest Service (USFS) *vs.* Bureau of Land Management (BLM) lands under different time periods. Estimates obtained from USFWS [43] data.

Federal Agency	Pre-Owl Listing (ha) (1981–1990)	Anticipated Rates (ha) (1991–2000)	Calculated Rates (1994–2003) (%)
USFS (WA, OR)	25,910	15,951	4,187 (0.21)
USFS (CA)	NA	1,903	669 (0.14)
BLM (OR)	8,907	9,474	1,988 (0.52)
Regional Total	NA	27,328	6,844 (0.24)

NA = not available.

Notably, extinction rates of spotted owls at the territory scale have been linked to the additive effects of decreased old-forest area and interference competition with barred owls [44]. Wiens *et al.* [39] also reported that the barred owl's competitive advantage over the spotted owl diminishes in spotted owl territories with a greater proportion of late-successional habitat. Thus, conservation of large tracts of contiguous, old-forest habitat is justified in any attempt to maintain northern spotted owls in the landscape.

Spotted Owls and Fires—USFWS [40] assumes that fire is a leading cause of habitat loss to owls on federal lands, However, few empirical studies have actually investigated northern spotted owl response to fire absent post-fire logging in or around owl territories [45,46]. Spotted owls may be resilient to forest fires provided low-moderate severity patches (refugia) are present within large fire complexes to provide nesting and roosting habitat. In the dry portions of the owls' range, where fire is common, owl fitness is associated with a mosaic of older forests (nesting and roosting habitat) and open vegetation patches (foraging areas; [47,48]). Such patch mosaics are produced by mixed-severity fires characteristic of the Klamath and eastern Cascade dry ecoprovinces [49,50] that may have contributed to maintenance of owl habitat historically [51]. However, if fire increases in severity or homogeneity of burn patterns due to climate change [52,53] and if LSOG losses outpace recruitment rates over time, the beneficial habitat effects of fire to owls would diminish. Currently, a deficit in high-severity fire exists in most of western North America compared to historical levels [49,54]. Recruitment of older forests in dry ecoprovinces of the region is projected to outpace fire losses for the next several decades [55].

Despite uncertainties about owl use of post-fire landscapes, federal managers in dry ecoprovinces have employed widespread forest thinning with the intent to reduce fire severity perceived as a threat to owl habitat. However, forest thinning may lead to cumulative losses in owl habitat that exceed those from severe fires. Using state transition models that accounted for recruitment of owl habitat over time *vs.* presumed habitat losses from severe fires, Odion *et al.* [55] concluded that thinning of suitable owl habitat at intensities (22% to 45% of dry forest provinces) recommended by USFWS [40] would reduce LSOG three to seven times more than loss attributed to high-severity fires. Projected thinning losses were consistent with empirically based studies of habitat loss from thinning that reduced overstory canopy below minimum thresholds for owl prey species [56]. The tradeoff between fire risk reduction and owl persistence in thinned forests has seldom if ever been systematically evaluated by the federal agencies.

2.4. Marbled Murrelet Continues to Decline but at a Slower Rate

Murrelet Population Trends—This federally threatened coastal seabird, nests in older-aged forests usually within 80-km of the coast from northern California to Alaska. The murrelet was listed as threatened in the Pacific Northwest due to habitat fragmentation from roads and clearcuts that expose murrelets to increased levels of nest predation [57–59]. Murrelet distribution and population trends are determined by the amount of suitable nesting habitat within five coastal "conservation zones" from Washington to California [60]. In general, as nesting habitat decreases murrelet abundance goes down, although abundance is also related to near-shore marine conditions (e.g., fish-prey abundance). Over the NWFP area, the trend estimate for the 2001–2013 period was slightly negative (~1.2%) (confidence intervals overlapped with zero [60]). At the scale of conservation zones, there was strong evidence of a linear decline in murrelet nesting populations in two of the five conservation zones both in Washington State. Trends were downward (but not significant) in other NWFP states; declines in murrelets likely would have been worse without the NWFP [60,61].

Murrelet Habitat Trends—About 1 million ha of potential suitable nesting habitat for murrelets remained on all lands within the range of the murrelet at the start of the NWFP (estimate based on satellite imagery [60]). Of this, only ~186,000 ha was estimated as high quality nesting habitat based on murrelet nest site locations. Over the NWFP baseline (1993–2012), net loss of potential nesting habitat was 2% and 27% on federal and nonfederal lands, respectively [60]. Losses on federal lands were mostly due to fire (66%) and logging (16%); on nonfederal lands logging (98%) was the primary cause of habitat loss [60]. In sum, loss and degradation of murrelet habitat resulted from: (1) logging on nonfederal lands (*i.e.*, State and private); (2) logging and thinning in suitable habitat and in habitat buffers on federal lands, including within LSRs; and (3) a variety of natural and anthropogenic causes including fire, windthrow, disturbance, and development [62].

Given that the availability of higher-quality nesting habitat is related to the carrying capacity of murrelets, forest management should focus on conserving and restoring remaining nesting habitat. The conservation strategy for murrelets, therefore, should include protecting remaining large patches of older-aged forests with minimal edge, buffering nest sites from windthrow and predators, and maintaining habitat connectivity. Maintaining the system of LSRs continues to be critical to murrelet conservation as is balancing the short- and long-term management of forests within LSRs [60,61]. For example, thinning that accelerates creation of older forest conditions in forest plantations that eventually become suitable to murrelet nesting can have short-term negative impacts, including increasing access of predators (e.g., corvids) to murrelet nest sites, blowdown and unraveling of suitable habitat, and changing the microclimate critical to temperature regulation and habitat availability [61]. Increased edge resulting from forest fragmentation can lower moss abundance needed for murrelet nesting [63,64], and increase nest depredation rates by corvids, especially at the juxtaposition of large openings and forests and in areas with berry producing plants such as elderberry (*Sambucus* sp. [65–67]. These factors underscore the need to maintain suitable buffers (suggested minimum widths of 91–183 m [57]) to minimize fragmentation and edge effects, and reduce windthrow and predation risk within

LSRs and adjacent to suitable murrelet habitat [60]. Landscape condition, juxtaposition of occupied murrelet habitat, and ownership should all be considered in thinning operations within LSRs or adjacent to older-aged forests.

Impacts to murrelets would increase if fire frequency and severity were to increase due to climate change. Greater storm intensity associated with climate change also may cause more windthrow, especially in fragmented landscapes. Because murrelet nesting and foraging habitat appear sensitive to climate variability [68], forest management for murrelets should consider the potential additive effects of climate change and habitat fragmentation. Maintaining the LSR network, protecting all occupied sites outside LSRs, and, in the long term, protecting all remaining habitat and minimizing fragmentation and edge effects are essential conservation measures [60–62].

2.5. The Aquatic Conservation Strategy Has Improved Watershed Conditions

The Aquatic Conservation Strategy of the NWFP established Riparian Reserves and Key Watersheds to restore and maintain ecological processes and the structural components of aquatic and riparian areas [69]. Protective stream buffers in Riparian Reserves preclude most logging and Key Watersheds are managed for water quality and habitat improvements for at-risk salmonids. Stream conditions across 214 watersheds are being evaluated on federal lands in two eight-year sampling periods (2002–2009 and 2010–2017, incomplete) [70].

At the regional scale, broad-scale improvements in pools (*i.e.*, deep water pockets that provide cover, food, thermal refuge for aquatic species), stream substrate, and aquatic macroinvertebrates were observed between sampling periods, but no trend was detected in physical habitat features in riparian area canopy cover condition or stream temperature (Table 3).

Table 3. Summary of aquatic trend analysis testing for linear relationship between sampling periods (2002–2009 and 2010–2013, incomplete) [69]. Macroinvertebrates were based on an observed to expected index (O/E) calculated by Miller *et al.* [69]. Pool scores were estimated by using the amount of fine (<2 mm) sediments that accumulate in the downstream portion of pools.

Aquatic Indicator	Trend Estimate	F-Test *	*p*-Value
Physical habitat	+0.1	0.33	0.59
Pools	−0.21	6.22	**0.03**
Wood	+0.09	3.14	0.11
Substrate	+0.10	9.90	**0.02**
Macro-invertebrates O/E	+0.01	10.84	**0.02**
Temperature	−0.09	1.19	0.31

* Includes Kenward-Roger approximation. $p < 0.05$ is significant as described in Miller *et al.* [69]

At the NWFP level, moderate gains in upslope/riparian conditions occurred due to forest ingrowth and road decommissioning; however, they were largely offset by declines in riparian forest cover following large fires, particularly in reserve areas [69]. Notably, the Aquatic Conservation Strategy anticipated that improvements in stream and habitat conditions would take place over many decades; repeated monitoring confirms short-term benefits as noted but long-term goals have yet to be realized [68,69,71]. With available data, watershed condition appeared best in Congressionally Reserved lands (primarily designated Wilderness Areas), followed by LSRs, and the Matrix, although statistical analysis could not be performed due to incomplete sampling [69]. Key Watersheds and roadless areas encompass many of the remaining areas of high-quality habitat and represent refugia for aquatic and riparian species [72]. Therefore, improved protection and restoration actions in those areas are critically important to conserving aquatic biodiversity. We note that in the smaller number of watersheds where riparian conditions have measurably declined in the past 25 years, largely due to wildfire, we can expect a pulsed, very rapid improvement of instream conditions in the coming decades. This is because of anticipated post-fire recruitment of large wood coupled with vigorous regrowth of vegetation in riparian areas and erosion-prone slopes—at least where these natural recovery processes have not been disrupted and delayed by post-fire logging.

In a recent review of the NWFP's Aquatic Conservation Strategy Frissell *et al* [73] documented a host of reasons to recommend expansion of Riparian Reserves, and reduction in logging compared to the original (baseline) NWFP. They recommended that Key Watersheds and LSRs receive more stringent protection to ensure their contribution to aquatic conservation and salmon recovery. They also called for more limits on or an end to post-fire logging, and more aggressive and strategically focused reduction of road density and storm proofing improvements in roads that remain. The BLM and the Forest Service, however, have increased logging in Riparian Reserves, are now proposing or suggesting reductions in the width and extent of Riparian Reserves, and have pressed for increasing road system density to provide access to more land for logging purposes. These agency recommendations do not explicitly consider ongoing stressors from land management in the surrounding nonfederal lands or increasing likelihood of climate-change-driven stress from drought, floods, and wildfire. Nor do they deal with the adverse watershed impacts from thinning projects relative to their putative but highly uncertain benefits for reducing the severity of future fire or insect outbreaks.

2.6. Climate Change and the NWFP

Climate change was not fully anticipated during development of the NWFP and thus represents a new broad-scale stressor that would exacerbate earlier projected and realized cumulative impacts to aquatic and terrestrial species and ecosystems throughout the region. Temperatures already have increased by 0.7 °C from 1895–2011 [53] and are anticipated to rise another 2 °C–6 °C by late century with warming most extreme during the summer [53,74]. Greater uncertainty exists in precipitation projections due to variability in emissions scenarios and climate models; however, summertime drying by the end of the century has higher certainty [53]. Summer drying coupled with increasing temperatures will likely impact timing of salmonid migrations in snow-fed streams [53,75] and increase future fire events [52,75].

Notably, a key characteristic of widely distributed species is that the dynamics of their multiple local populations experience environmental variation asynchronously. This decoupling of the dynamics of local populations within a metapopulation greatly increases overall persistence likelihood given inevitable large-scale disturbances [76]. Persistence is achieved because the spatial distribution of the species exceeds the spatial extent of most stochastic environmental events. Persistence may be compromised, however, when climate change operates as a top-down driver over very large spatial scales, increasing the synchrony of metapopulation dynamics and extinction probabilities for late-successional species. Persistence likelihood in the face of disturbances was addressed in the NWFP via redundancy and distribution of the reserve network but it is unclear whether the reserves can accommodate unprecedented climate-related shifts. This does not mean that the reserves are ineffective, just that they may not be as effective as hoped, and increasing the number and size of LSRs would make the network more effective.

Environmental uncertainty caused by climate change also has implications for restoration objectives of the NWFP. The NWFP assumed that young plantations can be restored to an older forest condition, but this may be less certain as forest succession comes under the influence of novel climatic conditions and perhaps increasingly altered disturbance regimes [52]. Thus, as forest conditions are altered by climate change, this may impact the climate preferences of late-successional species (e.g., mesic species are expected to decline near coastal areas due to drying [24]). One important way to reduce this uncertainty is to conserve more LSOG along north-facing slopes as potential micro-refugia and a hedge against further losses [24].

2.7. Ecosystem Services and the NWFP

Older forests and intact watersheds generally provide a myriad of ecosystem services associated with high levels of biodiversity [77,78]. Some examples of ecosystem services that have benefited from the NWFP include net primary productivity, water quality, recreation such as camping and hunting, salmon productivity, and carbon storage and sequestration. Older forests with high biomass (>200 mg carbon/ha, live above ground biomass of trees) most abundantly provide these services in aggregate primarily on federal lands [79].

The storage of carbon on federal lands is especially noteworthy because the region's high-biomass forests are among the world's most carbon dense forest ecosystems [80,81]. When cut down, these forests quickly release about half their carbon stores as CO_2 [82]. Reduced logging levels and increased regrowth under the NWFP has resulted in the regional forests shifting from a net source of CO_2 prior to the NWFP to a net sink for carbon during the NWFP time period [83]. While most of the carbon losses on federal lands are the result of forest fires, logging (mostly on nonfederal lands) remains the leading cause of land-use related CO_2 emissions [84]. Forests regenerating from natural disturbances including fire also rapidly sequester carbon and can then store it for long periods via succession if undisturbed. By comparison, logging places forests on short-rotation harvests, thereby precluding long-periods of carbon accumulation [82,83].

3. Building on the NWFP

The NWFP was founded on the best available science of the time, and the plan's reserve network and ecosystem management approach remain fundamentally sound [1,16,40,61,85,86] (also see http://www.fws.gov/oregonfwo/species/data/northernspottedowl/recovery/Plan/; accessed on 29 July 2015). If federal agencies wish to retain the protective elements of the NWFP, then forest plan revisions need to be based first and foremost on an adaptive approach to long-term goals as informed by monitoring. Increases in conservation measures are warranted to accommodate new scientific knowledge and unprecedented challenges from climate change and land-use stressors.

More recent climate change policies have been enacted since the NWFP that should be incorporated into forest planning. Examples include President Barack Obama's November 2013 Climate Change Executive Order directing federal agencies to include forest carbon sequestration in forest management, the Council on Environmental Quality's draft guidelines on reducing greenhouse gas emissions from land-used activities (Federal Register Vol. 80, No. 35/Monday, 23 February 2015), and emphasis on forest carbon and ecosystem integrity in forest planning on national forests [87]. Improvements to the NWFP's ecosystem and conservation focus are especially relevant today given: (1) the spotted owls' precarious status, including increased competition with barred owls; (2) continuing declines in murrelet populations; (3) other at-risk species recently proposed for listing (e.g., Pacific fisher *Martes pennanti*, North Oregon Coast Range distinct population segment of the red tree vole *Arborimus longicaudus*); (4) numerous forest associated invertebrates and lesser known species with restricted ranges that are vulnerable to extinction as a result of climate change [24]; and (5) additional ESU's of Pacific salmon that have been listed with none recovered to the point of delisting. Recent and ongoing land-use stressors acting alone or in concert, especially on nonfederal lands, also need to be reduced along with improved forest management practices and stepped up conservation efforts (Table 4).

Table 4. Land use stressors, the Northwest Forest Plan (current), and suggested additions based on adaptive management approaches.

Land Use Stressor	NWFP Current	Suggested NWFP Improvements
Climate-forced wildlife migrations	LSRs, landscape connectivity via riparian and other reserves	Enlarge LSR and riparian reserve network by protecting remaining older and high-biomass forests in the reserve system, increase connectivity for climate-forced wildlife displacement, reduce management stressors, shift older forests to the reserves and forest management to restoration of degraded areas, and identify and protect climate refugia [24], especially for rare and endemic species (continue the survey and manage program).
Livestock grazing	Aquatic Conservation Strategy standards and guidelines provide some protections for riparian and other sensitive areas	Remove cattle from riparian areas and reduce overall grazing pressure via large no-grazing zones given cumulative effects of grazing and climate change [88].
Wildfire	Thinning for fuels reduction and post-fire logging allowed in dry province reserves (trees <80 years) and Matrix	Prohibit post-fire logging in reserves, maintain all large snags in the Matrix (other than legitimate road side hazards), continue to protect older trees >80 years and maintain canopy closure at ≥60% in spotted owl habitat in thinning operations [55]. Plan for wildland fire to achieve ecosystem integrity objectives. Focus on flammable tree plantations and work cooperatively with private landowners on fire risk reduction.
Forest carbon loss	Not recognized other than if they overlap with reserves	Optimize carbon storage by protecting high-biomass forests from logging and by reducing logging frequency and intensity to sequester more carbon. Choose management alternatives with low emissions from forestry by making use of new assessment tools [89] (also see http://landcarb.forestry.oregonstate.edu/summary.aspx; accessed on 29 July 2015).
Aquatic ecosystem degradation	Riparian Reserves, Key Watersheds, LSRs, watershed restoration, watershed assessments/monitoring	Maintain or increase riparian buffer widths to ameliorate winter erosion, sedimentation, and flooding, restore floodplain connectivity and sinuosity, retain runoff and natural summer storage, increase efforts to improve and decommission failing roads, identify cold water refugia for increased protections [73,90], update watershed and LSR assessments to incorporate carbon and climate change. Where possible, support a closed forest canopy over perennial and intermittent streams and fully restore recruitment of large downed wood, including by prohibiting or severely limiting forest thinning in riparian reserves.

BLM Western Oregon Plan Revisions

A key contribution of the NWFP was its unprecedented emphasis on coordination among federal agencies via an overarching ecosystem management approach. In particular, the BLM manages ~1 million ha within the NWFP area (http://www.blm.gov/or/plans/wopr/oclands.php; accessed on 29 July 2015). BLM lands collectively provide irreplaceable ecosystem benefits to people and wildlife in western Oregon where there are relatively fewer national forest lands near the coast. Benefits include some 480,000 ha of watersheds that overlap with Surface Water Source Areas that produce clean drinking water for >1.5 million people from Medford to Portland, Oregon (State of Oregon water quality datasets; http://www.deq.state.or.us/wq/dwp/results.htm; accessed on 29 July 2015), connectivity and dispersal functions for wildlife linking the Coast and Cascade ranges (east-west, north-south linkages) [91], and habitat for at-risk species (Table 5). Unfortunately, the BLM has signaled its intent to move away from the Aquatic Conservation Strategy stream buffers and the survey and manage protections (http://www.blm.gov/or/plans/wopr/oclands.php; accessed on 29 July 2015).

Table 5. Summary of important ecological attributes of a subset of BLM lands in western Oregon essential to the coordinated management of the Northwest Forest Plan (summarized from Staus *et al.* [91]).

Attribute	BLM Lands
Late-successional forests	360,000 ha of old growth (>150 years, 22% of BLM Land), 236,000 ha mature (80–150 years, 15% of totals for western OR)
Northern spotted owl critical habitat	400,000 ha (27% of BLM land); LSRs: 240,000 ha
Marbled murrelet critical habitat	~192,000 ha, 32% of total critical habitat in western OR, 83% of which is within BLM LSRs
Evolutionary Significant Units of coho (*Oncorhynchus kisutch*)	~720,000 ha of coho ESU area, 260,000 ha of coho ESU's in BLM LSRs—35% of ESU area on BLM land. Of the 10,075 km of spawning and rearing habitat within western Oregon, 12% is located on BLM lands, 100% in Riparian Reserves, and 44% of which is within LSRs.
Evolutionary Significant Units of chinook (*O. tshawytscha*)	~148,000 ha of ESU habitat, 16% of BLM land in western Oregon; 25,200 ha of chinook ESU habitat in BLM LSRs—17% of the total ESU area on BLM land.
Evolutionary Significant Units of steelhead (*O. mykiss*)	87,200 ha of steelhead ESU habitat, all of which is found in the Salem and Eugene districts. Nine percent of BLM land in western Oregon contains steelhead ESU habitat with 14,000 ha of steelhead ESU habitat in BLM LSRs—16% of the total ESU area across BLM land.
Key Watersheds	Western Oregon contains ~1.6 million ha of Key Watersheds, 61,600 ha (4%) of which are located within BLM LSRs. In the Coast Range, LSRs protect 9% of Key Watersheds overall, over 25% of 10 of the 38 key watersheds in this area.
Survey and Manage Species	Of the 404 survey and manage species (primarily rare species at risk of local extirpation) recognized in the NWFP, 149 species are found on BLM land and 93 are found within BLM LSRs. LSRs in the Salem BLM District contain the highest concentration of these species (54), followed by Roseburg (39), and Coos Bay (35). Species include red tree vole (*Arborimus longicaudus*, an important food source for spotted owls), and many species of vascular plant, aquatic mollusk, lichen, fungi, and bryophyte.

4. Robust Conservation Additions to the NWFP

The NWFP provided a much-needed starting-place for a robust conservation strategy on federal forests in the face of climate change. For clarity, we organize our recommendations to improve the plan based on widely recognized principles of conservation biology and ecosystem management that also apply more broadly to large-landscape conservation planning.

4.1. Reserves

The large, well distributed, and redundant system of reserves was chosen based on specific requirements for the northern spotted owl that are still supported by the best available science [1,16,17,40–42,85]. At a minimum, we recommend continuation of the reserve network as a foundation for at-risk species in a changing climate and with increased stressors in the surrounding nonfederal lands. The NWFP reserves along with the survey and manage program function together as precautionary measures for species that are less mobile (e.g., many endemics) due to increasing stressors in the surroundings and climate change [19,24]. Given the redundancy and spacing requirements of the reserve system to address owl viability requirements, the network is likely to maintain older forest conditions over time by accommodating temporary losses from fire and other natural disturbances without compromising the integrity of the network [2,9], unless disturbances increase dramatically due to climate change [53]. The reserve system also is arranged along north-south gradients, including the Coast and Cascade ranges, elevation gradients, and topographically diverse areas, presumably allowing for climate-forced wildlife dispersal and climate refugia [24]. Large, contiguous federal ownerships and coordinated management of federal agencies under the standards and guidelines of the NWFP should continue to allow for adaptive responses to climatic change. Blocks of federal ownership also provide opportunities for wildland fire needed to restore and maintain ecosystem processes across a successional gradient [10,92,93].

The NWFPs' combination of coarse- and fine-filter approaches should continue to provide time for many wildlife to adjust and adapt to changing climatic conditions. Any effort to scale-back the reserves (as is currently being considered by federal agencies) must acknowledge that the NWFP architects aptly recognized that LSRs, Riparian Reserves, and Key Watersheds fit together in a cohesive manner to maintain long-term benefits to terrestrial and aquatic ecosystems. Reducing protections to reserves would create cumulative impacts across ecosystems. With new stressors like climate change and ongoing land-uses, reserve synergies and integrated strategies are even more important.

4.2. Forest Carbon

Regional carbon storage capacity can be increased if managers both protect carbon stores in older high biomass forests and allow young forests to re-grow for longer periods [83,84]. Managing for high-biomass forests is also associated with the multifunctionality of ecosystems because carbon dense forests are associated with high levels of biodiversity and numerous other ecosystem services [79]. Prudent management should integrate forest carbon policies with multiple use management objectives of federal agencies by optimizing carbon stored in older forests and extending timber harvest rotations to allow for longer periods of carbon sequestration and storage. Thus, forest managers can select management alternatives to minimize carbon flux from logging and land-uses by evaluating alternatives based on new carbon assessment tools (Table 4).

4.3. Aquatic Conservation

The variety of requirements for watershed analysis, reserve assessments, and monitoring under the NWFP has provided a foundation for tracking the plan's implementation objectives for aquatic ecosystems, at least at a regional scale. With improvements, aquatic ecosystem monitoring could provide integrated and sensitive indicators of ecosystem changes associated with climate shifts. Current Aquatic Conservation Strategy provisions, therefore, could be strengthened to help make aquatic ecosystems more resilient to climate change by (1) lessening cumulative watershed impacts particularly from the extensive road network on federal lands; (2) reducing the imprint of management disturbance on relatively high-integrity watersheds and roadless areas; (3) emphasizing maintenance of riparian areas, shade, floodplain processes, and recruitment of large wood from both near stream areas and unstable slopes; and (4) restoring migratory connectivity and fish passage to allow cold-water fisheries a better chance to occupy refugia less stressed by climate change.

4.4. At-Risk Species Recovery

Our understanding of threats to at-risk species has greatly advanced since passage of the Endangered Species Act (ESA) in 1973 and the NWFP in 1994. Specifically, the recognition that avoiding extinction is different than achieving recovery when it addresses the original ESA goal of " ... preserving the ecosystems upon which threatened and endangered species depend." Hence, implementation of the NWFP and enforcement of the ESA are linked objectives that together provide for the ecosystem and population needs of at-risk species among a host of other benefits.

To build on the complementarity of the NWFP and ESA, we recommend that at-risk species recovery (e.g., spotted owl, marbled murrelet, Pacific salmon) on federal lands include more habitat protections to reduce interactions with their competitors (e.g., spotted owls *vs.* barred owls), maintain genetic diversity [94], provide for resilient populations, and enable multiple local populations to be well-distributed throughout the NWFP area. Additionally, at least until land-use stressors are reduced, the survey and manage program should be continued to avoid the need for listing future at-risk species and expanded to include species that require complex early seral forests [10]. Managers can then select a broad suite of focal species that depend on all segments of successional gradient.

4.5. Adopting New Policies and Approaches

The foundation of the NWFP can be easily amended to accommodate new scientific information and elevated and novel stressors by building on its foundational elements (e.g., reserves, stream buffers, survey and manage). This can best be accomplished by incorporating recent national forest policies that emphasize ecosystem integrity [87] and climate change planning on federal lands (President Barack Obama's 2013 Climate Change Executive Order), reducing land-use stressors, and maintaining or restoring landscape connectivity to enable climate-forced wildlife migrations (Figure 3). Additionally, recent mapping of high-biomass forests [84] and carbon accounting in forestry practices (http://landcarb.forestry.oregonstate.edu/summary.aspx; accessed on 29 July 2015) provide new opportunities for retaining carbon in older forests while reducing forestry related CO_2 emissions.

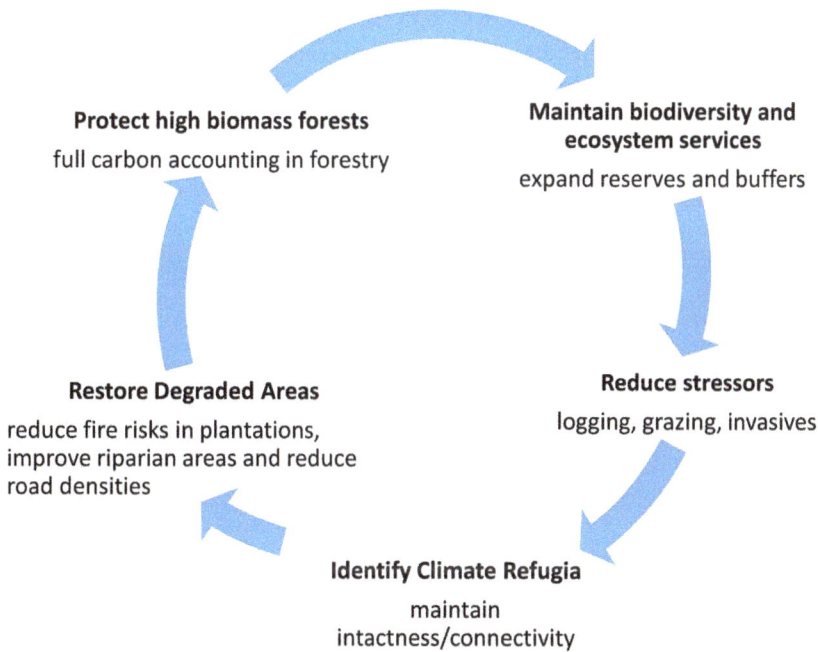

Protect high biomass forests
full carbon accounting in forestry

Maintain biodiversity and ecosystem services
expand reserves and buffers

Restore Degraded Areas
reduce fire risks in plantations, improve riparian areas and reduce road densities

Reduce stressors
logging, grazing, invasives

Identify Climate Refugia
maintain intactness/connectivity

Figure 3. Integrating ecosystem management and conservation biology with recent forest policies related to climate change (e.g., President Barack Obama's 2013 Climate Change Executive Order), forest carbon, and ecosystem integrity in forest planning [87].

5. Conclusions

The foundation of the NWFP is its reliance on best available science for conserving, restoring, and responsibly managing federal lands within the range of the northern spotted owl and, for the first time ever, an entire ecosystem, which is why it is considered a global model [1]. Although the plan is only two decades into its century-long implementation, its key conservation goals and species recovery mandates are far more likely to be met with the plan's management and conservation measures intact.

As forest plan revisions go forward in the region, the reserve network needs to be expanded in response to increasing land-use stressors to ecosystems and at-risk species, and to provide for a more robust conservation framework in response to climate change. Climate change may trigger more forest fires in places and, correspondingly, more logging and livestock grazing as these practices almost always follow forest fires on federal lands. Notably, burned forests successionally link complex early seral forests [10,11] to future old-forest development [92] and are not ecological disasters as often claimed. Depending on fire severity, burned forests provide nesting and roosting (low-moderate severity) or foraging (high severity) habitat for spotted owls [45,46,51]. Federal managers, however, have increasingly proposed massive post-fire logging projects that degrade complex early seral forests [95] and spotted owl habitat [45,46], and that can elevate fuel hazards and re-burn potential [96,97]. Post-fire logging over large landscapes may cause type conversions whereby fires burn intensely in logged areas only to be replanted in densely stocked and flammable tree plantations to burn intensely again in the next fire and so on [98]. Livestock grazing in combination with climate change is also now the biggest impact to biodiversity on federal lands that needs to be offset by new protections such as large blocks of ungrazed areas [88].

In sum, changes in ecosystem management practices on federal lands, triggered by the NWFP, have for the most part arrested an approaching ecosystem-wide collapse set in motion by decades of large-scale logging and mounting land-use stressors. Implementation of the plan has been challenging due, in large part, to socio-economic pressures to increase logging without full consideration of the environmental consequences and understanding of the science and conservation principles underpinning the NWFP. Moreover, despite substantive improvements in federal land management practices compared to those previous to the NWFP, amendments that respond to emerging contemporary threats are clearly needed. Scientific information and robust conservation principles can provide federal managers with the knowledge needed to adapt the next generation of forest plans. Improvements should be grounded in careful evaluation of the effects of past actions along with ongoing and future stressors as they pertain to the region's underlying ecological fabric and its link to sustainable economies. Science-based revisions of the plan should seek to improve its implementation in an adaptive context by addition rather than subtraction. Unfortunately, attempts to revise the plan have been bogged down by ongoing controversy over timber *vs.* biodiversity values that has led to a perpetual tug-of-war between decision makers that either support or seek to dismantle the NWFP. If this trend continues, federal land management may regress and recreate many of the problems the NWFP was implemented to correct, including re-inflamed social conflict, a cascade of endangered species listings, permanently increased conservation burdens on private landowners due to additional endangered species listings, and loss of ecological integrity that underpins the region's ecosystem services and their adaptive capacity to climate change.

Acknowledgments: This manuscript is dedicated to the late Robert G. Anthony whose decades of inspirational research and dedication to spotted owl recovery and role on FEMAT paved the way for the NWFP. C. Meslow provided constructive review of earlier versions of the manuscript and T. Spies provided input on old-growth forest communities. Wilburforce and Weeden foundations provided funding for D. DellaSala.

Conflicts of Interest: The authors declare no conflict of interest.

References

1. DellaSala, D.; Williams, J.E. Special Section: The Northwest Forest Plan: A global model of forest management in contentious times. *Conserv. Biol.* **2006**, *20*, 274–276. [PubMed]

2. Forest Ecosystem Management Assessment Team (FEMAT). *Forest Ecosystem Management: An Ecological, Economic, and Social Assessment*; USDA, US Department Interior Fish & Wildlife Service, US Department of Commerce, US Department of the Interior National Park Service, US Department Interior Bureau of Land Management, and Environmental Protection Agency: Portland, OR, USA, 1993.

3. *Record of Decision for Amendments to Forest Service and Bureau of Land Management Planning Documents within the Range of the Northern Spotted Owl*; April 13 1994. Available online: http://www.reo.gov/library/reports/newroda.pdf (accessed on 15 September 2015).

4. U.S. Fish and Wildlife Service (USFWS). *The 1990 Status Review: Northern Spotted Owl Strix Occidentalis Caurina*; U.S. Fish and Wildlife Service: Portland, OR, USA, 1990; p. 95.

5. Wimberly, M.C.; Spies, T.A.; Long, C.J.; Whitlock, C. Simulating historical variability in the amount of old forests in the Oregon Coast Range. *Conserv. Biol.* **2000**, *14*, 167–180. [CrossRef]

6. Old-growth Definition Task Group. *Interim Definitions for Old-Growth Douglas-Fir and Mixed-Conifer Forests in the Pacific Northwest and California*; Pacific Northwest Research Station, USDA Forest Service: Portland, OR, USA, 1986.

7. Franklin, J.F.; Spies, T.A. *Ecological Definitions of Old-Growth Douglas-Fir Forests*; General Technical Report PNW-GTR-85; U.S. Department of Agriculture Forest Service, Pacific Northwest Research Station: Portland, OR, USA, 1991.

8. Fierst, J. *Region 6 Interim Old Growth Definitions for Douglas-Fir Series, Grand Fir/White Fir Series, Interior Douglas-Fir Series, Lodgepole Pine Series, Pacific Silver Fir Series, Ponderosa Pine Series, Port-Orford-Cedar and tanoak (Redwood) Series, Subalpine Fir Series, and Western Hemlock Series*; U.S. Department of Agriculture Forest Service, Timber Management Group: Portland, OR, USA, 1993.

9. Davis, R.J.; Ohmann, J.L.; Kennedy, R.E.; Cohen, W.B.; Gregory, M.J.; Yang, Z.; Roberts, H.M.; Gray, A.N.; Spies, T.A. *Northwest Forest Plan–The First 20 Years (1994-2013): Status and Trends of Late-Successional and Old-Growth Forests (draft)*; USDA Forest Service: Portland, OR, USA, 2015. Available online: http://www.reo.gov/monitoring/reports/20yr-report/LSOG%2020yr%20Report%20-20Draft%20for%20web.pdf (accessed on 15 September 2015).

10. Swanson, M.E.; Franklin, J.F.; Beschta, R.L.; Crisafulli, C.M.; DellaSala, D.A.; Hutto, R.L.; Lindenmayer, D.B.; Swanson, F.J. The forgotten stage of forest succession: Early-successional ecosystems on forested sites. *Front. Ecol. Environ.* **2011**, *9*, 117–125. [CrossRef]

11. DellaSala, D.A.; Bond, M.L.; Hanson, C.T.; Hutto, R.L.; Odion, D.C. Complex early seral forests of the Sierra Nevada: What are they and how can they be managed for ecological integrity? *Nat. Areas J.* **2014**, *34*, 310–324. [CrossRef]

12. Strittholt, J.R.; DellaSala, D.A.; Jiang, H. Status of mature and old-growth forests in the Pacific Northwest, USA. *Conserv. Biol.* **2006**, *20*, 363–374. [CrossRef] [PubMed]

13. Grinspoon, E.; Jaworski, D.; Phillips, R. *Northwest Forest Plan-The First 20 Years (1994–2013) Socioeconomic Monitoring*; USDA Forest Service: Portland, OR, USA, 2015.

14. Charnley, S. The Northwest Forest Plan as a model for broad-scale ecosystem management: A social perspective. *Conserv. Biol.* **2006**, *20*, 330–340. [CrossRef] [PubMed]

15. Power, T.M. Public timber supply, market adjustments, and local economies: Economic assumptions of the Northwest Forest Plan. *Conserv. Biol.* **2006**, *20*, 341–350. [CrossRef] [PubMed]

16. Courtney, S.P.; Blakesley, J.A.; Bigley, R.E.; Cody, M.L.; Dumbacher, J.P.; Fleischer, R.C.; Franklin, A.B.; Franklin, J.F.; Gutiérrez, R.J.; Marzluff, J.M.; *et al.* *Scientific Evaluation of the Status of the Northern Spotted Owl*; Sustainable Ecosystems Institute: Portland, OR, USA, 2004. Available online: http://www.fws.gov/oregonfwo/species/data/northernspottedowl/BarredOwl/Documents/CourtneyEtAl2004.pdf (accessed on 15 September 2015).

17. Lint, J.B. *Northwest Forest Plan—The First 10 Years (1994–2003): Status and Trends of Northern Spotted Owl Populations and Habitat*; General Technical Report PNW-GTR-648; USDA Forest Service, PNW Research Station: Portland, OR, USA, 2005; p. 176.

18. Noon, B.R.; Murphy, D.; Beissinger, S.R.; Shaffer, M.L.; DellaSala, D.A. Conservation planning for US National Forests: Conducting comprehensive biodiversity assessments. *Bioscience* **2003**, *53*, 1217–1220. [CrossRef]

19. Carroll, C.; Odion, D.C.; Frissell, C.A.; DellaSala, D.A.; Noon, B.R.; Noss, R. *Conservation Implications of Coarse-Scale Versus Fine-Scale Management of Forest Ecosystems: Are Reserves Still Relevant?* Klamath Center for Conservation Research: Orleans, CA, USA, 2009; Available online: http://www.klamathconservation.org/docs/ForestPolicyReport.pdf (accessed on 17 September 2015).

20. Noss, R.F.; Dobson, A.P.; Baldwin, R.; Beier, P.; Davis, C.R.; DellaSala, D.A.; Francis, J.; Locke, H.; Nowak, K.; Lopez, R.; *et al.* Bolder thinking for conservation. *Conserv. Biol.* **2012**, *26*, 1–4. [CrossRef] [PubMed]

21. Watson, J.E.M.; Dudley, N.; Segan, D.B.; Hockings, M. The performance and potential of protected areas. *Nature* **2014**, *515*, 67–73. [CrossRef] [PubMed]

22. Molina, R.; Marcot, B.G.; Lesher, R. Protecting rare, old-growth, forest-associated species under the Northwest Forest Plan. *Conserv. Biol.* **2006**, *20*, 306–318. [CrossRef] [PubMed]

23. DellaSala, D.A.; Reid, S.B.; Frest, T.J.; Strittholt, J.R.; Olson, D.M. A global perspective on the biodiversity of the Klamath-Siskiyou ecoregion. *Nat. Areas J.* **1999**, *19*, 300–319.

24. Olson, D.M.; DellaSala, D.A.; Noss, R.F.; Strittholt, J.R.; Kaas, J.; Koopman, M.E.; Allnutt, T.F. Climate change refugia for biodiversity in the Klamath-Siskiyou ecoregion. *Nat. Areas J.* **2012**, *32*, 65–74. [CrossRef]

25. Thomas, J.W.; Forsman, E.D.; Lint, J.B.; Meslow, E.C.; Noon, B.R.; Verner, J. *A Conservation Strategy for the Northern Spotted Owl*; A Report of the Interagency Scientific Committee to Address the Conservation of the Northern Spotted Owl; U.S. Department of Agriculture Forest Service: Portland, OR, USA, 1990.

26. Boyce, M.S. Population viability analysis. *Annu. Rev. Ecol. Syst.* **1992**, *23*, 481–506. [CrossRef]

27. MacArthur, R.H.; Wilson, E.O. *Theory of Island Biogeography*; Princeton University Press: Princeton, NJ, USA, 1967.

28. Murphy, D.D.; Noon, B.R. Integrating scientific methods with habitat conservation planning: Reserve design for the Northern Spotted Owl. *Ecol. Appl.* **1992**, *2*, 3–17. [CrossRef]

29. Noon, B.R.; McKelvey, K.S. *A Common Framework for Conservation Planning: Linking Individual and Metapopulation Models*; McCullough, D.R., Ed.; Metapopulations and Wildlife Conservation, Island Press: Washington, DC, USA, 1996; pp. 139–166.

30. Noon, B.R.; McKelvey, K.S. Management of the Spotted Owl: A case history in conservation biology. *Annu. Rev. Ecol. Syst.* **1996**, *27*, 135–162. [CrossRef]

31. Den Boer, P.J. On the survival of populations in a heterogeneous and variable environment. *Oecologia* **1981**, *50*, 39–53. [CrossRef]

32. Goodman, D. Consideration of stochastic demography in the design and management of biological reserves. *Nat. Res. Model.* **1987**, *1*, 205–234.

33. Lande, R. Risks of population extinction from demographic and environmental stochasticity, and random catastrophes. *Am. Nat.* **1993**, *142*, 911–927. [CrossRef]

34. Brown, J.H.; Kodric-Brown, A. Turnover rates in insular biogeography: Effect of immigration on extinction. *Ecology* **1977**, *58*, 445–449. [CrossRef]

35. Fahrig, L.; Merriam, G. Habitat patch connectivity and population survival. *Ecology* **1985**, *66*, 1762–1768. [CrossRef]

36. Noon, B.R.; Lamberson, R.H.; Boyce, M.S.; Irwin, L.L. Population viability analysis: A primer on its principal technical concepts. In *Ecological Stewardship: A Common Reference for Ecosystem Management*; Szaro, R.C., Johnson, N.C., Eds.; Elsevier Science: New York, NY, USA, 1999; Volume 2, pp. 87–134.

37. Davis, R.; Falxa, G.; Grinspoon, E.; Haris, G.; Lanigan, S.; Moeur, M.; Mohoric, S. *Northwest Forest Plan: The First 15 Years (1994–2008)*; R6-RPM-TP-03-2011; USDA Forest Service, Pacific Northwest Research Station: Portland, OR, USA, 2011.

38. Forsman, E.D.; Anthony, R.G.; Dugger, K.M.; Glenn, E.M.; Franklin, A.B.; White, G.C.; Schwarz, C.J.; Burnham, K.P.; Anderson, D.R.; Nichols, J.D.; *et al. Population Demography of Northern Spotted Owls. Studies in Avian Biology 40*; University of California Press: Berkley, CA, USA, 2011; p. 106.

39. Wiens, J.D.; Anthony, R.G.; Forsman, E.D. Competitive interactions and resource partitioning between northern spotted owls and barred owls in Western Oregon. *Wildl. Monogr.* **2014**, *185*, 1–50. [CrossRef]

40. U.S. Fish & Wildlife Service (USFWS). *Revised Recovery Plan for the Northern Spotted Owl (Strix Occidentalis Caurina)*; USDI Fish and Wildlife Service: Portland, OR, USA, 2011.

41. Dugger, K.; Andrews, S.; Brooks, J.; Burnett, T.; Fleigel, E.; Friar, L.; Phillips, T.; Tippin, T. *Demographic Characteristics and Ecology of Northern Spotted Owls (Strix Occidentalis Caurina) in the Southern Oregon Cascades*; Annual Research Report; Oregon Cooperative Fish and Wildlife Research Unit (OCFWRU), Department of Fisheries and Wildlife, Oregon State University: Corvallis, OR, USA; p. 24.

42. Anthony, R.G.; Forsman, E.D.; Franklin, A.B.; Anderson, D.R.; Burnham, K.P.; White, G.C.; Schwarz, C.J.; Nichols, D.J.; Hines, J.E.; Olson, G.; *et al.* Status and trends in demography of northern spotted owls, 1985–2003. *Wildl. Monogra.* **2006**, *163*, 1–48. [CrossRef]

43. U.S. Fish and Wildlife Service. (USFWS). *2007 Draft Recovery Plan for the Northern Spotted Owl (Strix Occidentalis Caurina) Merged Options 1 and 2*; USFWS: Portland, OR, USA, 2007.

44. Dugger, K.M.; Anthony, R.G.; Andrews, L.S. Transient dynamics of invasive competition: Barred owls, spotted owls, habitat, and the demons of competition present. *Ecol. Appl.* **2011**, *21*, 2459–2468. [CrossRef] [PubMed]

45. Clark, D.A.; Anthony, R.G.; Andrews, L.S. Survival rates of northern spotted owls in post-fire landscapes of southwest Oregon. *J. Raptor Res.* **2011**, *45*, 38–47. [CrossRef]

46. Clark, D.A.; Anthony, R.G.; Andrews, L.S. Relationship between wildfire, salvage logging, and occupancy of nesting territories by northern spotted owls. *J. Wildl. Manag.* **2013**, *77*, 672–688. [CrossRef]

47. Franklin, A.B.; Anderson, D.R.; Gutiérrez, R.J.; Burnham, K.P. Climate, habitat quality, and fitness in Northern Spotted Owl populations in northwestern California. *Ecol. Monogr.* **2000**, *70*, 539–590. [CrossRef]

48. Dugger, K.M.; Wagner, F.; Anthony, R.G.; Olson, G.S. The relationship between habitat characteristics and demographic performance of northern spotted owls in southern Oregon. *Condor* **2005**, *107*, 865–880. [CrossRef]

49. Hanson, C.T.; Odion, D.C.; DellaSala, D.A.; Baker, W.L. Overestimation of fire risk in the Northern Spotted Owl recovery plan. *Conserv. Biol.* **2009**, *23*, 1314–1319. [CrossRef] [PubMed]

50. Odion, D.C.; Hanson, C.T.; Arsenault, A.; Baker, W.L.; DellaSala, D.A.; Hutto, R.L.; Klenner, W.; Moritz, M.A.; Sherriff, R.L.; Veblen, T.T.; *et al.* Examining historical and current mixed-severity fireregimes in ponderosa pine and mixed-conifer forests of western North America. *PLoS ONE* **2014**, *9*, 1–14.

51. Baker, W.L. Historical Northern Spotted Owl habitat and old-growth dry forests maintained by mixed-severity wildfires. *Landscape Ecol.* **2015**, *30*, 665–666. [CrossRef]

52. Littell, J.S.; McKenzie, D.; Peterson, D.L.; Westerling, A.L. Climate and wildfire area burned in western U.S. ecoprovinces, 1916–2003. *Ecol. Appl.* **2009**, *19*, 1003–1021. [CrossRef] [PubMed]

53. Mote, P.; Snover, A.K.; Capalbo, S.; Eigenbrode, S.D.; Glick, P.; Littell, J.; Raymondi, R.; Reeder, S. North-west. In *Climate Change Impacts in the United States: The Third National Climate Assessment*; Melillo, J.M., Richmond, T.C., Eds.; U.S. Global Change Research Program: Washington, DC, USA, 2014; pp. 487–513. [CrossRef]

54. Baker, W.L. Are high-severity fires burning at much higher rates recently than historically in dry-forest landscapes of the western USA? *PLoS ONE* **2015**, *10*, e0136147. [CrossRef] [PubMed]

55. Odion, D.C.; Hanson, C.T.; DellaSala, D.A.; Baker, W.L.; Bond, M.L. Effects of fire and commercial thinning on future habitat of the northern spotted owl. *Open Ecol. J.* **2014**, *7*, 37–51. [CrossRef]

56. DellaSala, D.A.; Anthony, R.G.; Bond, M.L.; Fernandez, E.; Hanson, C.T.; Hutto, R.L.; Spivak, R. Alternative views of a restoration framework for federal forests in the Pacific Northwest. *J. For.* **2013**, *111*, 402–492. [CrossRef]

57. U.S. Fish & Wildlife Service (USFWS). *Recovery Plan for the Threatened Marbled Murrelet (Brachyramphus Marmoratus) in Washington, Oregon, and California*; USDI Fish and Wildlife Service: Portland, OR, USA, 1997; p. 203.

58. Luginbuhl, J.M.; Marzluff, J.M.; Bradley, J.E.; Raphael, M.G.; Varland, D.E. Corvid Survey techniques and the relationship between corvid relative abundance and nest predation. *J. Field Ornithol.* **2001**, *72*, 556–572. [CrossRef]

59. Lynch, D.; Roberts, L.; Falxa, G.; Brown, R.; Tuerler, B.; D'Elia, J. *Evaluation Report for the 5-Year Status Review of the Marbled Murrelet in Washington, Oregon, and California*; Unpublished report; Fish and Wildlife Service, Region 1: Lacey, WA, USA, 2009.

60. Falxa, G.A.; Raphael, M.G. *Northwest Forest Plan—The First 20 Years (1994–2013): Status and Trend of Marbled Murrelet Populations and Nesting Habitat (Draft)*; U.S. Department of Agriculture, Forest Service, Pacific Northwest Research Station: Portland, OR, USA, 2015.

61. Raphael, M. Conservation of the marbled murrelet under the Northwest Forest Plan. *Conserv. Biol.* **2006**, *20*, 297–305. [CrossRef] [PubMed]

62. McShane, C.; Hamer, T.; Carter, H.; Swartzman, G.; Friesen, V.; Ainley, D.; Tressler, R.; Nelson, K.; Burger, A.; Spear, L.; *et al.* *Evaluation Report for the 5-Year Status Review of the Marbled Murrelet in Washington, Oregon, and California*; Unpublished report; EDAW, Inc.: Seattle, WA, USA; Prepared for the U.S. Fish and Wildlife Service, Region 1: Portland, OR, USA, 2004.

63. Malt, J.; Lank, D. Temporal dynamics of edge effects on nest predation risk for the marbled murrelet. *Biol. Conserv.* **2007**, *140*, 160–173. [CrossRef]

64. Van Rooyen, J.C.; Malt, J.M.; Lank, D.B. Relating microclimate to epiphyte availability: Edge effects on nesting habitat availability for the marbled murrelet. *Northwest Sci.* **2011**, *85*, 549–561. [CrossRef]

65. Masselink, M.N.M. Responses of Steller's Jays to Forest Fragmentation on Southwest Vancouver Island and Potential Impacts on Marbled Murrelets. Master's Thesis, Department of Biology, University of Victoria, Victoria, BC, Canada, 2001; p. 138.

66. Marzluff, J.M.; Millspaugh, J.J.; Hurvitz, P.; Handcock, M.S. Relating resources to a probabilistic measure of space use: Forest fragments and Steller's Jays. *Ecology* **2004**, *85*, 1411–1427.

67. Marzluff, J.M.; Neatherlin, E. Corvid responses to human settlements and campgrounds: Causes, consequences, and challenges for conservation. *Biol. Conserv.* **2006**, *130*, 301–314. [CrossRef]

68. Becker, B.H.; Peery, M.Z.; Beissinger, S.R. Ocean climate and prey availability affect the trophic level and reproductive success of the marbled murrelet, and endangered seabird. *Mar. Ecol. Prog. Ser.* **2007**, *329*, 267–279. [CrossRef]

69. Reeves, G.H.; Williams, J.E.; Burnett, K.M.; Gallo, K. The aquatic conservation strategy of the Northwest Forest Plan. *Conserv. Biol.* **2006**, *14*, 319–329. [CrossRef]

70. Miller, S.A.; Gordon, S.N.; Eldred, P.; Beloin, R.M.; Wilcox, S.; Raggon, M.; Andersen, H.; Muldoon, A. *Northwest Forest Plan–The First 20 Years (1994-2013): Watershed Condition Status and Trend (draft)*; U.S. Department of Agriculture, Forest Service, Pacific Northwest Research Station: Portland, OR, USA, 2015.

71. Gallo, K.; Lanigan, S.H.; Eldred, P.; Gordon, S.N.; Moyer, C. *Northwest Forest Plan—The First 10 Years (1994–2003): Preliminary Assessment of the Condition of Watersheds*; General Technical Report PNW-GTR-647; USDA Forest Service, Pacific Northwest Research Station: Portland, OR, USA, 2005; p. 133.

72. Lanigan, S.H.; Gordon, S.N.; Eldred, P.; Isley, M.; Wilcox, S.; Moyer, C.; Andersen, H. *Northwest Forest Plan—The First 15 Years (1994–2008): Status and Trend of Watershed Condition*; General Technical Report PNW-GTR-856; USDA Forest Service, Pacific Northwest Research Station: Portland, OR, USA, 2012.

73. Frissell, C.A.; Baker, R.J.; DellaSala, D.A.; Hughes, R.M.; Karr, J.R.; McCullough, D.A.; Nawa, R.K.; Rhodes, J.; Scurlock, M.C.; Wissmar, R.C. *Conservation of Aquatic and Fishery Resources in the Pacific Northwest: Implications of New Science for the Aquatic Conservation Strategy of the Northwest Forest Plan*; Report prepared for the Coast Range Association: Corvallis, OR, USA, 2014; p. 35. Available online: http://coastrange.org (accessed on 29 July 2015).

74. DellaSala, D.A.; Brandt, P.; Koopman, M.; Leonard, J.; Meisch, C.; Herzog, P.; Alaback, P.; Goldstein, M.I.; Jovan, S.; MacKinnon, A.; *et al.* Climate Change may Trigger Broad Shifts in North America's Pacific Coastal Rainforests. In *Reference Module in Earth Systems and Environmental Sciences*; Elsevier: Boston, MA, USA, 2015; Available online: http://dx.doi.org/10.1016/B978-0-12-409548-9.09367-2 (accessed on 15 September 2015).

75. Littell, J.S.; Elsner, M.M.; Binder, L.C.W.; Snover, A.K. *The Washington Climate Change Impacts Assessment: Evaluating Washington's Future in a Changing Climate)*; Climate Impacts Group, University of Washington: Seattle, WA, USA, 2009.

76. Hanski, I. A practical model of metapopulation dynamics. *J. Anim. Ecol.* **1994**, *63*, 151–162. [CrossRef]

77. Hector, A.; Bagchi, R. Biodiversity and ecosystem multifunctionality. *Nature* **2007**, *448*, 188–190. [CrossRef] [PubMed]

78. Gamfeldt, L.; Snall, T.; Bagchi, R.; Jonsson, M.; Gustafsson, L.; Kjellander, P.; Ruiz-Jaen, M.C.; Froberg, M.; Stendahl, J.; Philipson, C.D.; *et al.* Higher levels of multiple ecosystem services are found in forests with more tree species. *Nat. Commun.* **2013**, *4*, 1340. [CrossRef] [PubMed]

79. Brandt, P.; Abson, D.J.; DellaSala, D.A.; Feller, R.; von Wehrden, H. Multifunctionality and biodiversity: Ecosystem services in temperate rainforests of the Pacific Northwest, USA. *Biol. Conserv.* **2014**, *169*, 362–371. [CrossRef]

80. Smithwick, E.A.H.; Harmon, M.E.; Remillard, S.M.; Acker, S.A.; Franklin, J.F. Potential upper bounds of carbon stores in forests of the Pacific Northwest. *Ecol. Appl.* **2002**, *12*, 1303–1317. [CrossRef]

81. Keith, H.; Mackey, B.G.; Lindenmayer, D.L. Re-evaluation of forest biomass carbon stocks and lessons from the world's most carbon-dense forests. *Proc. Natl. Acad. Sci. USA* **2009**, *106*, 11635–11640. [CrossRef] [PubMed]

82. Harmon, M.E.; Ferrell, W.K.; Franklin, J.F. Effects on carbon storage of conservation of old-growth forests to young forests. *Sci. Febr.* **1990**, *247*, 4943.

83. Krankina, O.N.; Harmon, M.E.; Schnekenburger, F.; Sierra, C.A. Carbon balance on federal forest lands of western Oregon and Washington: The impact of the Northwest Forest Plan. *For. Ecol. Manag.* **2012**, *286*, 171–182. [CrossRef]

84. Krankina, O.; DellaSala, D.A.; Leonard, J.; Yatskov, M. High biomass forests of the Pacific Northwest: Who manages them and how much is protected? *Environ. Manag.* **2014**, *54*, 112–121. [CrossRef] [PubMed]

85. Noon, B.R.; Blakesley, J.A. Conservation of the northern spotted owl under the Northwest Forest Plan. *Conser. Biol.* **2006**, *20*, 288–296. [CrossRef]

86. U.S. Fish & Wildlife Service (USFWS). *Revised Critical Habitat for the Northern Spotted Owl*; USDI Fish and Wildlife Service: Portland, OR, USA, 2012.

87. USDA Forest Service. *Final Programmatic Environmental Impact Statement National Forest System Land Management Planning*; USDA Forest Service: Washington, DC, USA, 2012. Available online: http://www.fs.usda.gov/planningrule (accessed on 17 September 2015).

88. Beschta, R.L.; DellaSala, D.A.; Donahue, D.L.; Rhodes, J.J.; Karr, J.R.; O'Brien, M.H.; Fleishcner, T.L.; Deacon-Williams, C. Adapting to climate change on western public lands: Addressing the impacts of domestic, wild and feral ungulates. *Environ. Manag.* **2013**, *53*, 474–491.

89. Sonne, E. Greenhouse gas emissions from forestry operations: A life cycle assessment. *J. Environ. Qual.* **2006**, *35*, 1439–1450. [CrossRef] [PubMed]

90. Isaak, D.J.; Young, M.K.; Nagel, D.E.; Horan, D.L.; Groce, M.C. The cold-water climate shield: Delineating refugia for preserving salmonid fishes through the 21st century. *Glob. Chang. Biol.* **2015**. [CrossRef] [PubMed]

91. Staus, N.L.; Strittholt, J.R.; DellaSala, D.A. Evaluating areas of high conservation value in western Oregon with a decision-support model. *Conserv. Biol.* **2010**, *24*, 711–720. [CrossRef] [PubMed]

92. Donato, D.C.; Campbell, J.L.; Franklin, J.F. Multiple successional pathways and precocity in forest development: Can some forests be born complex? *J. Veg. Sci.* **2012**, *23*, 576–584. [CrossRef]

93. DellaSala, D.A.; Hanson, C.T. Preface. In *The Ecological Importance of Mixed-Severity Fires:Nature's Phoenix*; DellaSala, D.A., Hanson, C.T., Eds.; Elsevier: Boston, MA, USA, 2015; pp. xxiii–xxxviii.

94. Funk, C.W.; Forsman, E.D.; Johnson, M.; Mullins, T.D.; Haig, S.M. Evidence for recent population bottlenecks in northern spotted owls (*Strix Occidentalis Caurina*). *Conserv. Genet.* **2009**, *11*, 1013–1021. [CrossRef]

95. Beschta, R.L.; Rhodes, J.J.; Kauffman, J.B.; Gresswell, R.E.; Minshall, G.W.; Karr, J.R.; Perry, D.A.; Hauer, F.R.; Frissell, C.A. Postfire management on federal public lands of the western United States. *Conserv. Biol.* **2004**, *18*, 957–967. [CrossRef]

96. Donato, D.C.; Fontaine, J.B.; Campbell, J.L.; Robinson, W.D.; Kauffman, J.B.; Law, B.E. Post-wildfire logging hinders regeneration and increases fire risk. *Science* **2006**, *311*, 352. [CrossRef] [PubMed]

97. Thompson, J.R.; Spies, T.A.; Garuio, L.M. Reburn severity in managed and unmanaged vegetation in a large wildfire. *Natl. Acad. Sci. USA* **2007**, *104*, 10743–10748. [CrossRef] [PubMed]

98. DellaSala, D.A.; Lindenmayer, D.B.; Hanson, C.T.; Furnish, J. In the aftermath of fire: Logging and related actions degraded mixed and high-severity burn areas. In *The Ecological Importance of Mixed-Severity Fires: Nature's Phoenix*; DellaSala, D.A., Hanson, C.T., Eds.; Elsevier: Boston, MA, USA, 2015; pp. 313–343.

forests

Correction

Correction: DellaSala, D.A., *et al.* Building on Two Decades of Ecosystem Management and Biodiversity Conservation under the Northwest Forest Plan, USA. *Forests*, 2015, 6, 3326

Dominick A. DellaSala [1,*], **Rowan Baker** [2], **Doug Heiken** [3], **Chris A. Frissell** [4], **James R. Karr** [5], **S. Kim Nelson** [6], **Barry R. Noon** [7], **David Olson** [8] **and James Strittholt** [9]

1 Geos Institute, 84-4th Street, Ashland, OR 97520, USA
2 Independent Consultant, 2879 Southeast Kelly Street, Portland, OR 97202, USA; watershedfishbio@yahoo.com
3 Oregon Wild, P.O. Box 11648, Eugene, OR 97440, USA; dh@oregonwild.org
4 Flathead Lake Biological Station, 32125 Bio Station Lane, University of Montana, Polson, MT 59860-6815, USA; leakinmywaders@yahoo.com
5 102 Galaxy View Court, Sequim, WA 98382, USA; jrkarr@u.washington.edu
6 Department of Fisheries and Wildlife, Oregon State University, 104 Nash Hall, Corvallis, OR 97331-3803, USA; kim.nelson@oregonstate.edu
7 Department of Fish, Wildlife and Conservation Biology, Colorado State University, Fort Collins, CO 80523, USA; barry.noon@colostate.edu
8 Conservation Earth Consulting; 4234 McFarlane Avenue, Burbank, CA 91505, USA; conservationearth@live.com
9 Conservation Biology Institute, 136 SW Washington Ave #202, Corvallis, OR 97333, USA; stritt@consbio.org
* Correspondence: dominick@geosinstitute.org; Tel.: +1-541-482-4459 (ext. 302); Fax: +1-541-482-4878

Academic Editor: Diana F. Tomback
Received: 19 February 2016; Accepted: 23 February 2016; Published: 26 February 2016

We discovered two typos and a change in a sentence needed in our published manuscript. Regarding the typos: timber estimates in Section 2 of the NWFP's Long-Term Objectives should have been 2.34 million cubic meters annually instead of 234 million cubic meters annually and ~1.78 million cubic meters instead of 178 million cubic meters annually. The statement regarding murrelet populations—"Trends were downward (but not significant) in other NWFP states" should be deleted in this paper [1]. The authors would like to apologize for any inconvenience caused to the readers by these changes.

References

1. DellaSala, D.A.; Baker, R.; Heiken, D.; Frissell, C.A.; Karr, J.R.; Nelson, S.K.; Noon, B.R.; Olson, D.; Strittholt, J. Building on Two Decades of Ecosystem Management and Biodiversity Conservation under the Northwest Forest Plan, USA. *Forests* **2015**, *6*, 3326–3352. [CrossRef]

MDPI

St. Alban-Anlage 66

4052 Basel

Switzerland

Tel. +41 61 683 77 34

Fax +41 61 302 89 18

www.mdpi.com

Forests Editorial Office

E-mail: forests@mdpi.com

www.mdpi.com/journal/forests

www.ingramcontent.com/pod-product-compliance
Lightning Source LLC
Chambersburg PA
CBHW051314020426
42333CB00028B/3342